Thomas Wright

The Ruins of the Roman City of Uriconium, at Wroxeter, near Shrewsbury

Thomas Wright

The Ruins of the Roman City of Uriconium, at Wroxeter, near Shrewsbury

ISBN/EAN: 9783744782173

Printed in Europe, USA, Canada, Australia, Japan

Cover: Foto ©berggeist007 / pixelio.de

More available books at **www.hansebooks.com**

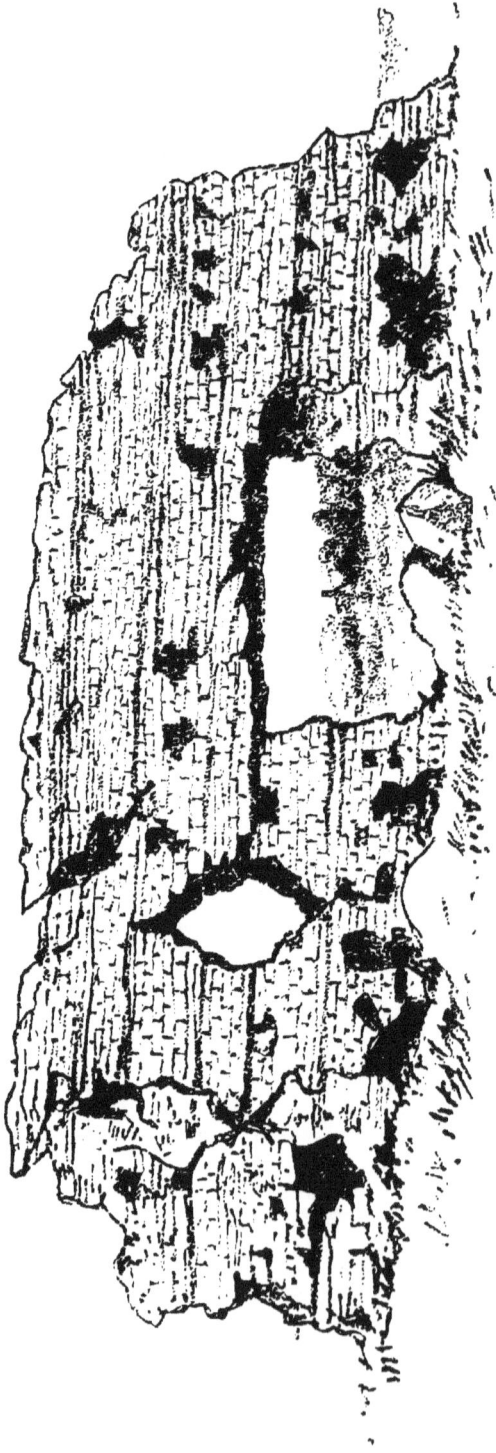

THE NORTH SIDE OF THE OLD WALL, AT WROXETER

RUINS OF THE ROMAN CITY

OF

URICONIUM.

BY

THOMAS WRIGHT, ESQ: M.A. F.S.A.

Wroxeter Church.

SHREWSBURY. BUNNY & EVANS; HIGH STREET
1860.

THE

RUINS OF THE ROMAN CITY

OF

Uriconium,

AT

WROXETER, NEAR SHREWSBURY.

BY

THOMAS WRIGHT, ESQ., M.A., F.S.A.

Sixth Edition,
WITH ILLUSTRATIONS.

SHREWSBURY: BUNNY & EVANS, HIGH-STREET.
LONDON: SIMPKIN, MARSHALL, & CO., PATERNOSTER-ROW.

1877.

TO VISITORS.

Parties from a distance, wishing to visit the ruins of the ancient Uriconium, at Wroxeter, will find every comfort and accommodation at the Raven, the Lion, the George, and the Crown Hotels, Shrewsbury.

Wroxeter is a little more than five miles from Shrewsbury. Conveyances may be obtained at the Railway Station, Shrewsbury, and at any of the Hotels or Livery Stables. Parties of any reasonable number may be conveyed by either of the latter, on giving them two days' notice by letter.

Upton Magna, on the Shrewsbury and Wellington Joint Line of Railway, is the nearest station to Wroxeter, from whence it is distant about two miles and a half— a pleasant walk for an active person.

The Museum of the Shropshire and North Wales Natural History and Antiquarian Society, College Hill, where all the moveable articles from Wroxeter are deposited, is open DAILY to visitors, from 10 till 4, on payment of threepence each, and by an order from a Subscriber GRATUITOUSLY.

List of Plates.

PREFACE.

———

It is the aim of the following pages to give the degree and kind of popular information believed to be wanted by the numerous visitors to the excavations at Wroxeter, who have no guide to explain what they see, and are not possessed of that amount of minute antiquarian knowledge which would enable them to understand everything without such explanation. It is the first instance in which there has been, in this country, the chance of penetrating into a city of more than fourteen centuries ago, on so large a scale, and with such extensive remains of its former condition; and when the visitor has walked over the floors which had been trodden last, before they were thus uncovered, by the Roman inhabitants of this island, he will appreciate more justly, and with greater interest, the objects which have been discovered, and are deposited in the Museum at Shrewsbury; and he will learn to look forward with hope to the light which a continuance of these excavations must throw upon the condition and history of this country at so remote a period. Whatever this light may be, it must not be forgotten that we shall

be indebted for it, in the first place, to His
Grace the Duke of Cleveland, who has shown
a generous public feeling in giving permission
and encouragement to the excavations on his
land, and to the late B. Botfield, Esq., M.P.,
through whose zeal and liberality in the
undertaking the excavators were set at work,
when as yet it was uncertain if their labours
would be attended with any success. I have
endeavoured to fulfil literally the title of this
little book, and to give the visitor such
information as he would seek from a pro-
fessional guide, whilst I have gladly left the
description of the Museum, and especially of
those rather numerous human remains which
form so remarkable a part of our discoveries,
to one best qualified for that task, Dr. Henry
Johnson, who has so ably and zealously di-
rected the excavations on the spot, and who
has thus, unremunerated, given to the service
of the public so much of his valuable time.

T. W.

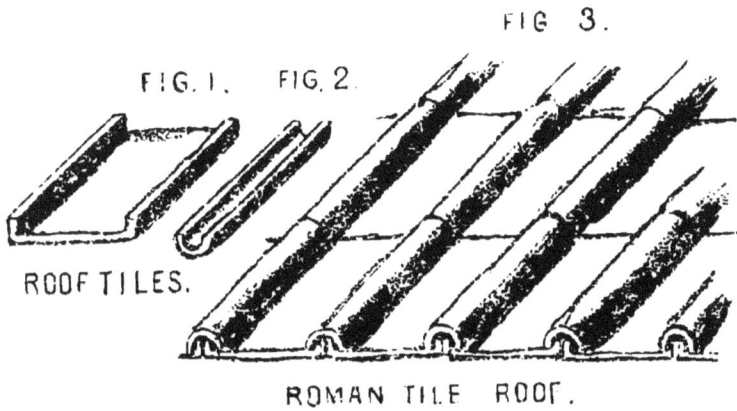

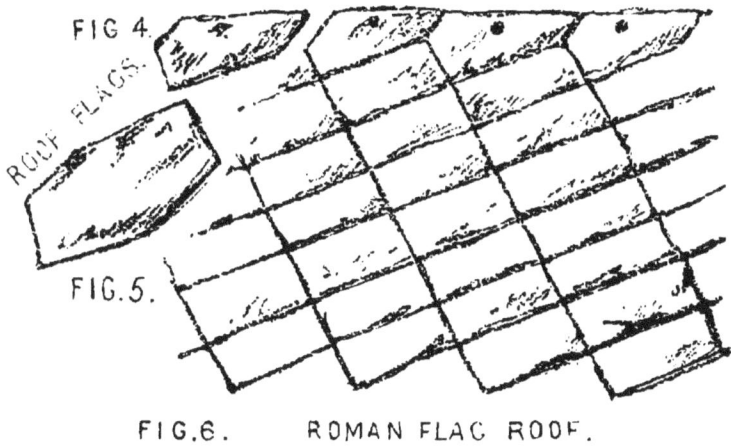

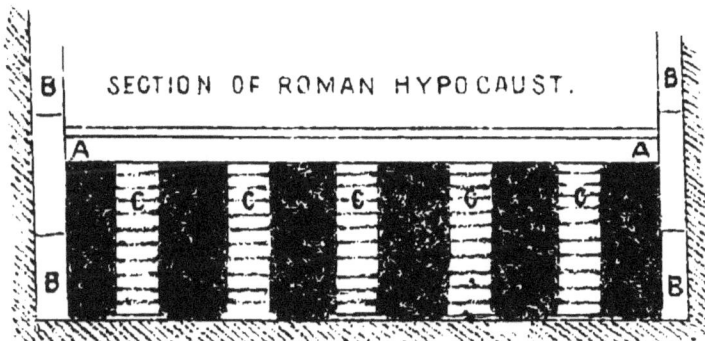

PLATE . 4.

FIG 3.

FIG. 1.   FIG. 2.

ROOF TILES.

ROMAN TILE ROOF.

FIG 4.
ROOF FLAGS.

FIG. 5.

FIG. 6.   ROMAN FLAG ROOF.

SECTION OF ROMAN HYPOCAUST.

FIG 7.

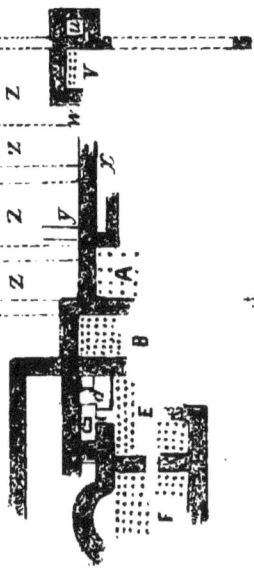

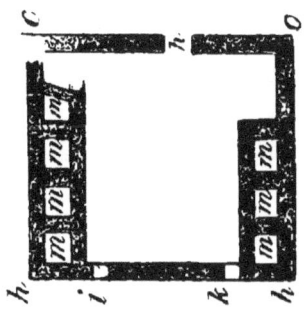

Plan of Walls,
discovered by excavations.
Uroxeter. Salop.
1859.

WALLS MARKED THUS XXX ARE COVERED IN

a WALL STILL ABOVE GROUND

SCALE OF   20   0   30   40   60   80   100
FEET

PLATE 5.

PLATE 7.

ANCIENT STONE FONT.
IN WROXETER CHURCH.

FIG 1

FIG 2

FIG 3.

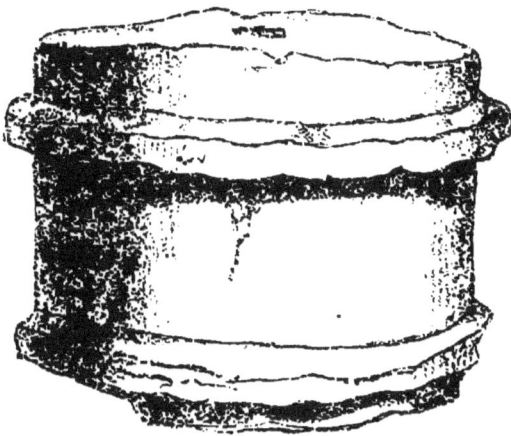

CAPITALS FOUND AT URICONIUM.

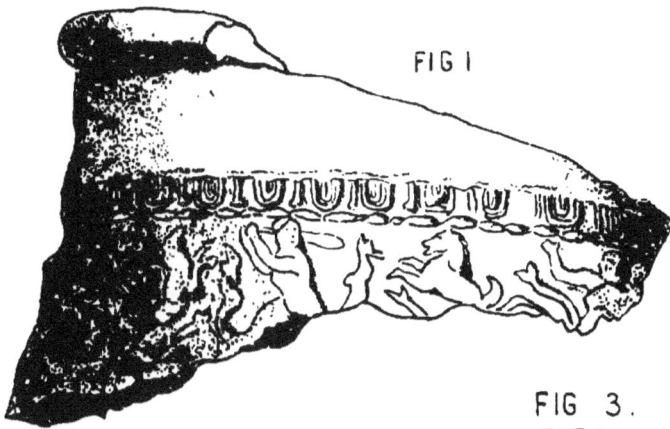

FIG 1

FIG 2.

FIG 3.

FIG 4.

FIG 5.

FIG.1, FRAGMENT OF SAMIAN WARE.
FIG.2, CINERARY URN.
FIGS.3,4,5, ROMANO SALOPIAN POTTERY.

FIG . 1 .

FIG. 2

FIG. 4

FIG. 3 .

FIG. 5 .

FIG . 6 .

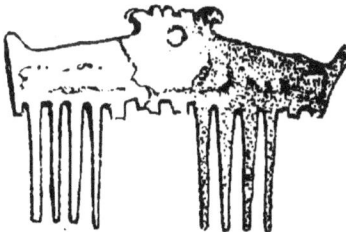

RINGS AND COMBS. (ACTUAL SIZE.)

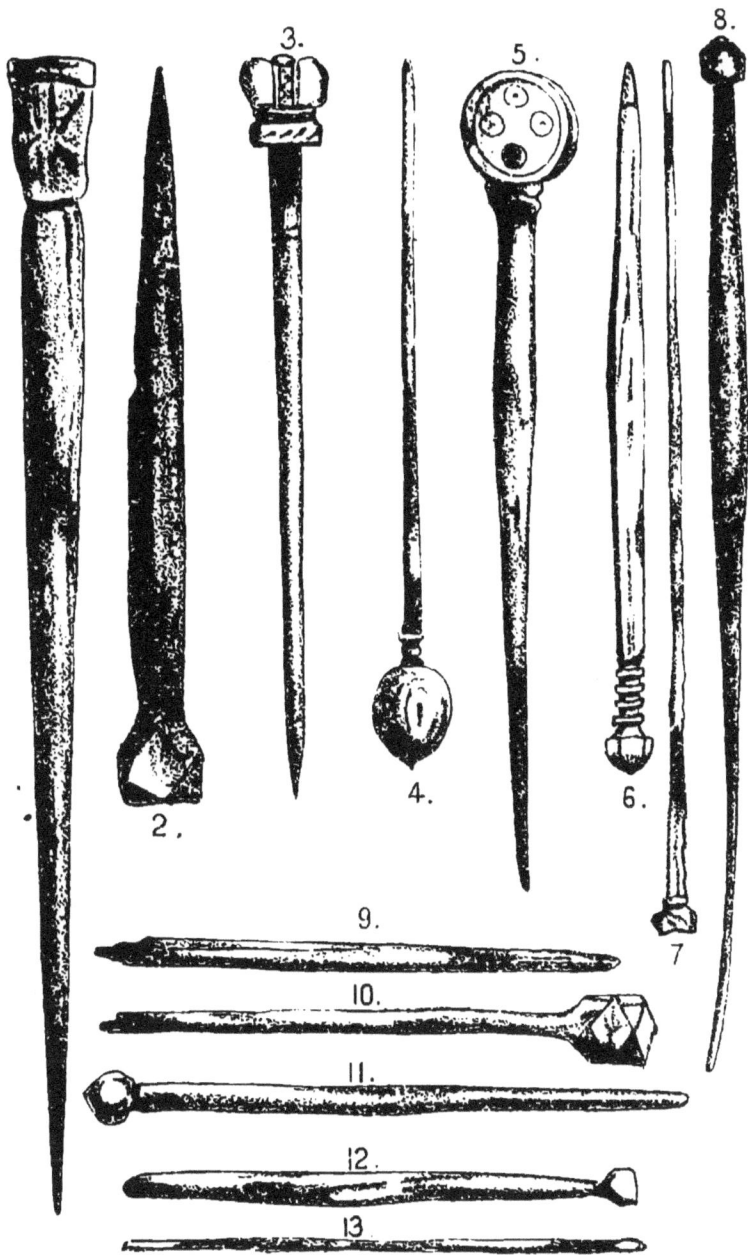

HAIR PINS, ETC. (ACTUAL SIZE.)

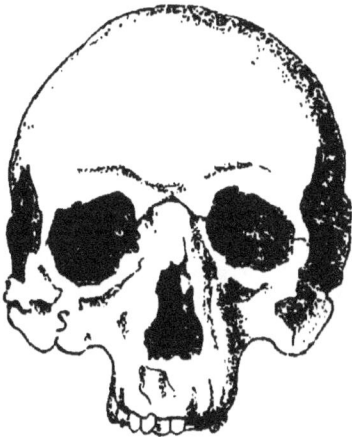

X.

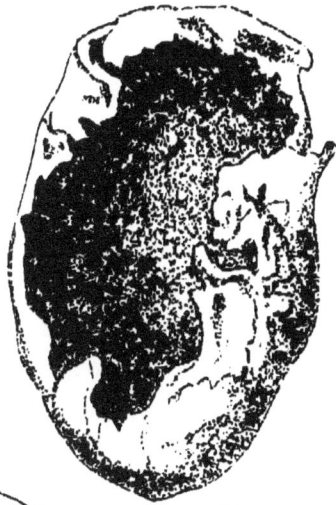

IV.

X.

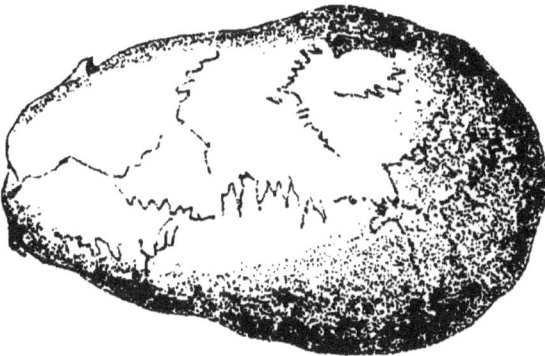

X.

SKULLS FROM WROXETER.

FIG. 1

FIG 2.

FIG 3

FIG 5

FIG 4.

FIG 5 MASK (ACTUAL SIZE) IN THE SHREWSBURY MUSEUM.
FIGS 1,2,3,& 4, ROMAN REMAINS FROM WROXETER, IN THE
POSSESSION OF SAMUEL WOOD, ESQ, SHREWSBURY.

PLATE 13.

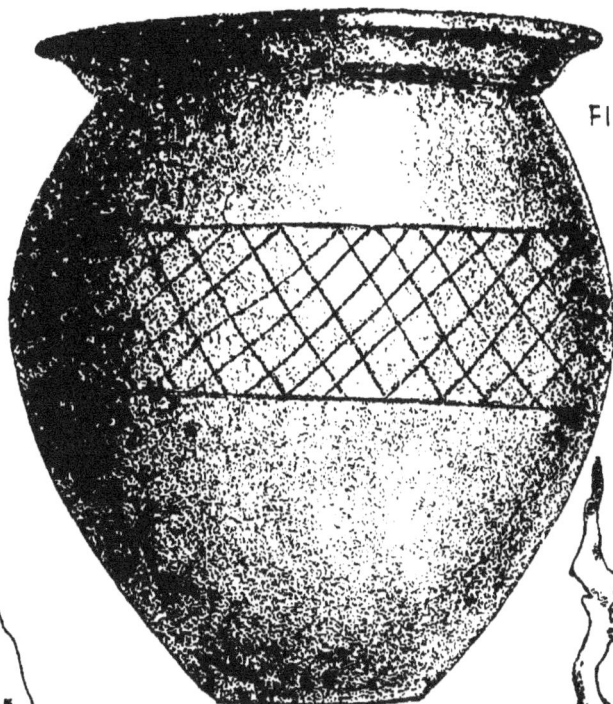

FIG 2.

FIG 4.

FIG 1.

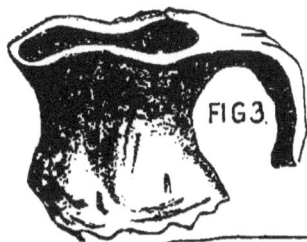

FIG 3.

FIG 5.

FIGS 2,& 3, UPCHURCH POTTERY.
FIG 4. ADZ.
FIG 5, ROMAN SALOPIAN (RED) WARE.
FIG 1, SPEAR HEAD.

FIG 1

FIG 2

SCALE

12 Inches

CARVED STONE FRAGMENTS.
FROM URICONIUM.
IN THE GARDEN OF EDWARD STANIER ESQ
WROXETER.

RUINS OF URICONIUM.

IF we leave Shrewsbury by its long eastern suburb, known, from the important monastic house which formerly stood at its commencement, as the Abbey Foregate, passing the more modern monument erected at its extremity, Lord Hill's Column, our way lies for about two miles along the London Road, bounded on each side by rich and fertile fields. At the distance just mentioned, this road approaches close to the river Severn, and continues to run along its banks, to the great improvement of the scenery, until we arrive at the prettily-situated village of Atcham, with Atcham Church in face of us, and the river winding under its stone bridge in the foreground. Atcham is three miles from Shrewsbury. Crossing the bridge, we leave the river, which here takes a long sweep to the southward, and follow the

B

road, which skirts for more than half a mile the extensive park of Attingham. We here approach another river, the Tern, which at this point spreads into a fair expanse of water, and forms, with the mansion of Attingham to the left, and the copses which skirt it, a scene of striking beauty, while to the right it divides into two branches which empty themselves into the Severn, a little lower down. Crossing Tern Bridge, and proceeding a short distance, still skirting the park, we reach a point where, opposite the entrance to Attingham Park, a branch road turns off to the right from the old London road. We must take this branch road which will lead us to the village of Wroxeter. We soon cross a small stream, which is known by the name of the Bell Brook, and after we have passed this brook, the visitor will hardly fail to remark, wherever his eye rests upon ploughed ground, the extraordinary blackness of the soil in comparison with that of the land over which he has previously passed.

In fact he has now entered upon the site of an ancient Roman city, which is known, from the circumstance of its being mentioned by the geographer Ptolemy, to have been standing here as early as the beginning of the second century, when it was called Viroconium,—a name which appears to have been changed in the later Romano-British period to Uriconium; at least this is the form under which the name occurs in the later geographers, and which has been generally adopted by modern antiquaries.

From the point at which we have now arrived,
the line of the ancient town-wall may be traced
by a continuous low mound, which runs southward towards the Severn, the banks of which
it follows for some distance, and, after passing
between the river and the modern village of
Wroxeter, turns eastwardly behind the vicarage-
house, and makes a long sweep till it reaches
the hamlet of Norton to the north, whence it
turns to the westward again, and reaches the
point from which we started, forming an
irregular oval, rather more than three miles
in circumference. A portion of the Bell Brook
runs through the Roman city. After crossing
this brook, we approach ground which rises
gently, and nearly at the highest point we see
to the left a smith's shop. At this spot, which
is rather more than five miles from Shrewsbury,
the road which has brought us from that town
crosses another road, which turns down to the
right, to the village of Wroxeter, not quite half
a mile distant. Wroxeter is an Anglo-Saxon
name, the first part of which is probably cor-
rupted from that of the ancient Roman city
of the site of which it occupies the southern
extremity. The road which has led us to it
is called the Watling Street Road, and there
is every reason for believing that it occupies
in a part of its course the line of one of the
principal streets of Uriconium. It crosses the
river Severn immediately below the village,
where there was doubtless a bridge in Roman
times, for it is in the highest degree improbable

that in approaching a town of such importance,
the Romans would cross a river like the Severn
only by a ford.

On arriving at the smith's shop just
alluded to, the attention of the visitor will
be attracted by a solid mass of masonry, which
forms a very imposing object, and presents
those unmistakable characteristics of Roman
work,—the long string-courses of large flat red
bricks. This mass of masonry, the only portion
of the buildings of Uriconium which remains
standing above ground, is upwards of twenty
feet high, and seventy-two feet long, with a
uniform thickness of three feet, and has been
long known by the name of "The Old Wall."
It stands nearly in the centre of the ancient
city, which occupied the highest ground within
the walls,—a commanding position, with the
bold isolated form of the Wrekin in the rear,
and in front a panorama of mountains formed
by the Wenlock and Stretton Hills, Caer Cara-
doc, the Longmynd, the Breidden, and the still
more distant mountains of Wales. With the
exception of this wall, all that remained of the
Roman city, if as some people might perhaps
have doubted, anything did remain,—has been
long buried beneath the soil. At the close of
the year 1858, however, it was resolved to
ascertain what these remains were, and an
Excavation Committee was formed at Shrewsbury, for the purpose of carrying this design
into effect by means of a public subscription.
Excavations were, accordingly, commenced on

the 3rd of February, 1859, and they have already led to results of the most satisfactory description. But, perhaps, before we proceed to describe the ruins which have thus been uncovered, it would be well to tell our readers something of the general character of the Roman towns in this island, and to explain how some of them were destroyed, and from what causes and by what circumstances their remains present themselves in the conditions in which we now find them.

FIFTEEN hundred years ago, this island, with the exception of the highlands of Scotland, was covered with flourishing towns, many of them known to have been of considerable magnitude, situated on numerous public roads,— these latter of such excellent construction, that they have remained to the present day the foundation of most of our great English high roads. These towns, like those in other parts of the empire, enjoyed free municipal institutions (from which our own mediaeval municipal institutions are derived), and in all but certain duties towards the imperial government, formed in themselves so many little republics, possessing all the ambitions and rivalries which

seem inseparable from republican institutions.
Among the slight notices of this island in
ancient writers we learn that the towns of
Britain were remarkable for their turbulence,
which was encouraged, no doubt, by the distance of this province from Rome, and by the
peculiar character of the population of the
towns, which consisted of blood that was
foreign to the soil, and which was not uniform
in character in the different towns. We know
further that, during the fourth century, those
towns often confederated together, threw off the
imperial yoke, and raised emperors of their own ;
and we have every reason for supposing that,
when the restraint imposed by the central
power became slackened, the towns confederated against one another, and that domestic
dissensions and contests troubled the peace of
the island. Such dissensions left the island
exposed to the invasions of its foreign enemies,
which had become very frequent and very
formidable during the fourth century. The
eastern coasts were often visited by the Teutonic rovers, Saxons, and Franks ; the barbarous
Caledonians, then called Picts, from the north
rushed across the borders, and carried devas-
tation through the land, in which they were
assisted by the Irish, or, as they were then
called, Scots, and probably by the Armorican
Celts, or Britons from Gaul. The towns of
Britain united would, no doubt, have presented
a force sufficient to meet any of these invasions, but their very constitution rendered

such a union difficult, except for a short period. Besides their independence of each other, the towns had only been expected to defend themselves, while the defence of the province was more especially the duty of the legions, and on their withdrawal, the towns seem to have followed their old practice in case of invasion, and shut themselves within their walls, or, at most, opposed the invaders without any union, thus leaving the open country to easy destruction.

The history of the conquest of the Roman provinces by the barbarians is, in general, simply the successive reduction of one town after another. Such was eminently the case in Britain, and the traditionary annals of the early Saxon period present little more than a list of conquered towns. Sometimes a town was taken by stratagem or force, and then it was plundered or destroyed, but in the far greater number of cases the town was too strong for the assailants and only submitted by composition, by which it paid a tribute to the conqueror and retained its old independent municipal institutions. We all know how many of our old cities and early municipal towns are thus the representatives of the cities of the Romans. In some parts of the island the destruction was greater than in others, and on the Welsh border, through the whole space between Chester (called by the Romans Deva), and Gloucester (which the Romans called Glevum,) the towns seem to have been

all ruined. One of the largest of these towns
was no doubt that of Uriconium. We can only
judge by implication, and by a comparison of
what occurred in other places, of the manner
in which a town like Uriconium was treated,
when it was overcome by the barbarians. We
know that these invaders were influenced by
a love of plunder, but a love of destruction—
we may perhaps call it an impulse of destruction—was still greater; and it is probable
that the plundering of a town like Uriconium
was a hasty and imperfect operation, and that
the plunderers carried off chiefly objects made
of the precious metals, or articles of dress
and arms, or other objects on which they
set considerable value, as they moved about
rapidly, and could not be provided very extensively with the means of conveyance. (We
are here speaking of the earlier plundering
invasions of the barbarians, such as the Picts
and Scots, in which perhaps Uriconium perished,
towards the middle of the fifth century, for it
is hardly probable that the Angles or Saxons
could have reached this part of the island at
so early a period.) The first impulse of the
plunderers was to apply fire to the buildings,
and the progress of the conflagration would
hasten their departure. Where the inhabitants
of the conquered town had not made their
escape and abandoned it before it was taken—
which was perhaps the case in some of the
smaller towns—there would no doubt be a
dreadful massacre, and the survivors would be

dragged away into captivity, for the various peoples who preyed upon the carcass of the mighty empire of Rome, whether German or Celt, or Tartar or Arab, ambitioned, almost above other plunder, the possession of numerous slaves. Thus the plundered town was left without inhabitants, and in flames, of which the latter, as may be judged, on the spot from the massive character of the walls of the houses, were probably partial in their effect, destroying chiefly the timber and roofs.

Thus the town was left an extensive mass of blackened walls; and such was the condition in which the ruined Roman towns remained during several centuries. Roman walls, we all know, were too strongly built to fall down, and various circumstances combined for their preservation. In the first place, the population of the country must have been greatly reduced, and this part of the island especially was probably very thinly inhabited after it had been ravaged by the invaders. The ruins themselves would in time be overgrown with plants and trees, and would become the haunt of wild beasts, which were then abundant, thus offering very little encouragement to anybody to enter them. But they were protected in a still greater degree by the strong superstitious feelings with which such ruins were regarded by the people who now occupied the land. The Teutonic invaders had not only a prejudice against towns in general, but they believed that all the deserted buildings

of the previous lords of the soil were taken
possession of by powerful evil spirits, on whose
limits it was in the highest degree dangerous
to trespass. They imagined, moreover, that
the Romans had the power of casting spells
over buildings, which were no less dangerous
than the evil spirits themselves. It will be
remembered how, when Augustine and his
brother missionaries came to convert the Anglo-
Saxons to Christianity, the Kentish king and
his court gave them their first audience in
the open air; because, as we are told, the
Anglo-Saxons were afraid that, should they be
received in a covered chamber in the palace,
the strangers from Rome would be able to cast
a spell upon them. It is a remarkable proof
of the strength of this superstitious feeling,
that all the Benedictionals of the Anglo-Saxon
period contain forms for blessing the vessels
of metal or earthenware found in ancient sites,
and relieving them from the spells which had
been cast upon them by the "pagans," in order
that the finders might be enabled to make use
of these vessels without any personal danger.
When the people of the middle ages, whether
Christians or not, found the beautiful bronze
figures on which we set so much store, they
were in the greatest apprehension of personal
danger until they had mutilated them so as to
break the charm or spell which they believed to
be laid upon them, for they looked upon these
images as the more general instruments of the
ancient magicians. When thus mutilated they

usually threw them into the nearest river. The numerous bronzes dredged up from the bed of the Thames at London are almost all mutilated in this manner. This was the case also with the inscriptions, for the successors of the Romans had no other notion of an ancient inscription than that it was a magical charm. This superstition has continued to exist until very recent times, for it appears that, within the memory of man, the peasantry of Northumberland, on the line of the great wall of Hadrian, were accustomed, when they found an inscribed stone—and inscribed stones are there very abundant—to hew out at least a part of the letters of the inscription with a pick or axe, in order to destroy the charm.

We thus understand how a ruined city—like that at Wroxeter—was allowed to remain untouched for centuries. Many of these ruined towns became the subject of romantic legends. One of these legends relating to an ancient ruined city in this. neighbourhood, is told in the curious History of the Fitz-Warines, composed in the thirteenth century, in Anglo-Norman, no doubt by a border writer. This writer is describing a visit supposed to have been made by William the Conqueror to the Welsh border in order to distribute the land to his followers. "When King William approached the hills and valleys of Wales he saw a very large town, formerly enclosed with high walls, which was all burnt and ruined, and in a plain below the town he caused his

tents to be raised, and there he said he would
remain that night. Then the king inquired
of a Briton what was the name of the town,
and how it came to be so ruined. 'Sire,' said
the Briton, 'I will tell you. The Castle was
formerly called Castle Bran, but now it is
called the Old March. Formerly there came
into this country Brutus, a very valiant knight,
and Corineus, from whom Cornwall still retains
its name, and many others derived from the
lineage of Troy, and none inhabited these parts
except very foul people, great giants, whose
king was called Geomagog. These heard of
the arrival of Brutus, and sent out to encounter
him, and at last all the giants were killed
except Geomagog, who was marvellously great.
Corineus, the valiant, said that he would will-
ingly wrestle with Geomagog, to try Geomagog's
strength. The giant, on the first onset, em-
braced Corineus so tightly, that he broke three
of his ribs. Corineus became angry, and struck
Geomagog with his foot that he fell from a
great rock into the sea, and Geomagog was
drowned. And a spirit of the devil now
entered into the body of Geomagog, and came
into these parts, and held possession of the
country long, that never Briton dared to
inhabit it. And long afterwards, King Bran
the son of Donwal, caused the city to be rebuilt, repaired the walls, and strengthened the
great fosses, and he made Burgh and Great
March. And the devil came by night and
took away every thing that was therein, since

which time nobody has ever inhabited there.'
The king marvelled much at this story, and
Payn Peverel, the proud and courageous knight,
the king's cousin, heard it all, and declared
that that night he would essay the marvel.
Payn Peverel armed himself very richly, and
took his shield, shining with gold, with a
cross of azure indented, and fifteen knights
and other attendants, and went into the highest
palace, and took up his lodging there. And
when it was night the weather became so foul,
black, dark, and such a tempest of lightning
and thunder, that all those who were there
became so terrified that they could not for
fear move hand or foot, but lay on the ground
like dead men. The proud Payn was very
much frightened, but he put his trust in God,
whose sign of the cross he carried with him,
and saw that he could have no help but from
God. He lay upon the ground, and with good
devotion prayed God and his mother Mary,
that they would defend him that night from
the power of the devil. Hardly had he
finished his prayer, when the fiend came in
the semblance of Geomagog, and he carried a
great club in his hand, and from his mouth
cast fire and smoke, with which the whole
town was illuminated. Payn had a good hope
in God, and signed himself with a cross, and
boldly attacked the fiend. The fiend raised
his club and would have struck Payn, but he
avoided the blow. The devil, by virtue of the
cross, was all struck with fear, and lost his

strength, for he could not approach the cross.
Payn pursued him till he struck him with
his sword; then he began to cry out, and fell
flat to the ground, and yielded himself vanquished. 'Knight,' said he, 'you have conquered me, not by your own strength, but by
virtue of the cross which you carry.' 'Tell me,'
said Payn, 'you foul creature, who you are, and
what you do in this town, I conjure you, in
the name of God and of the Holy Cross.'
The fiend began to relate from word to word
as the Briton had said before; and told how,
when Geomagog was dead, he immediately
rendered his soul to Beelzebub, their prince,
and he entered the body of Geomagog, and
came in his semblance into these parts, and
kept the great treasure which Geomagog had
collected and put into a house he had made
underground in that town. Payn demanded
of him, 'what kind of creature he was?' and
he said, 'He was formerly an angel, but now
is, by his forfeit, a diabolical spirit.' 'What
treasure,' said Payn, 'had Geomagog?' 'Oxen,
cows, swans, peacocks, horses, and all other
animals made of fine gold; and there was a
golden bull, which, through me, was his prophet,
and in him was all his belief; and he told
him the events that were to come; and twice
a year the giants used to honour their god,
the golden bull, whereby so much gold is
collected that all this country was called 'The
White Land.' And I and my companion inclosed the land with a high wall and deep

fosse, so that there was no entrance except
through this town, which was full of evil
spirits.' 'Now, you shall tell me,' said Payn,
'where is the treasure of which you have
spoken?' 'Vassal,' said he, 'speak no more
of that, for it is destined for others; but you
shall be lord of all this honour.'" And so
the vanquished fiend goes on to tell him the
future fortunes of his house; and after King
William had been duly informed of this ad-
venture, and they had thrown the body of
Geomagog into a great pit, they proceeded
on their way to Oswestry.

In my edition of this History of the Fitz-
Warines I have offered some conjectures on
the spot to which this legend refers; but on
comparing all the circumstances connected with
it, I have since been led to the conclusion that
the "burnt and ruined" city which had thus
been taken possession of by the evil spirits
was no other than the ruins of the ancient
Uriconium. This story implies that the walls
of the town and houses of Uriconium were still
standing above ground as late as the eleventh
and twelfth centuries, and very likely a great
portion of them remained thus standing at
the time when the author of the "History of
the Fitz-Warnes" wrote. But during the
centuries which had passed since the city of
the Romans became a ruin, it had been under-
going a gradual but continual change from the
accumulation of earth. This rising of the level
of the ground is always found to have taken

place under such circumstances, and may be explained by several causes. In the first place, the floors must have been covered by a mass of rubbish formed by the falling in of the roofs and more perishable parts of the buildings. Vegetation, too, would in the course of years arise, and the walls would stop and cause to be deposited the dust and earthy particles carried about in the atmosphere. This deposit we know by experience to be considerable. It is now little more than three centuries since the dissolution of the monasteries, and we have all had opportunities of observing the depth of earth under which the floors of the monastic ruins now lie, sometimes amounting to as much as three or four feet. What, then, must it have been on an extensive ruin like that of Uriconium, which had stood in that ruined and deserted condition from the middle of the fifth century to the middle of the twelfth?

It was at this latter period that the Roman buildings began to be systematically destroyed. It appears that still in the twelfth century, England was covered with the remains of Roman ruined towns and villas standing above ground, as they are still seen, though on a larger scale, in the countries which formed the Roman province in Northern Africa. We have seen the superstitious feelings which prevented people approaching those ruins in our island, and it required nothing less than the hand of the Church to interfere and break the charm which kept the rest of society aloof.

We learn from the history of the abbots of St.
Alban's, written in the thirteenth century by
Matthew Paris, that already in the eleventh
century the abbots of that great religious house
had began to break the ruins of the Roman city
of Verulamium, in order to use them as building materials. This practice became very
general in the twelfth century, and from that
time the Roman ruins were pillaged on an
extensive scale whenever a monastery or a
church was to be built. The ancient city at
Wroxeter was probably one of the great quarries
from which the builders of Haughmond Abbey
were supplied, and no doubt it contributed
materials to other monastic houses in this part
of the country. The church of Atcham, the
adjoining parish, and that of Wroxeter itself,
bear evidence to this appropriation of building
materials taken from ancient Uriconium. At
the time when this inroad was made upon the
ruins, the ground, as explained before, was
already raised several feet above the Roman
floors; and the mediæval builders, finding
plenty of material above ground, cleared away
the walls down to the surface of the ground as
it then existed, and sought them no further.
This accounts for the condition in which we
now find these walls, that is, remaining tolerably
perfect just up to the height of what was the
level of the ground, at the time the rest was
destroyed. The difference between the tops of
the walls as they now exist underground, and
the present surface of the ground, is the accu-

c

mulation of earth which has taken place since
this destruction.   It was the destruction of the
buildings which first caused this accumulation,
by scattering about the fragments of the plaster
of the walls and the broken tiles and stones
which were not worth carrying away. After
the walls above ground disappeared, and the
ground was levelled and cleared, such accumulation went on much more slowly.

The sites of the ancient towns, thus
cleared, and the spell which held their invaders
at bay having been broken by the ecclesiastics,
became exposed to a new class of depredators.
Coins and objects of some value were no doubt
discovered from time to time by accident, and
were greatly exaggerated by common report,
during ages when the existence of hidden
treasure formed a prominent article in the
popular belief. Many a Salopian, doubtless,
longed for the hidden treasures of the city of
Geomagog, and many an attempt no doubt was
made to discover and obtain them. Treasure-
hunting of this description was a great pursuit
with our mediæval forefathers, and the same
superstitious feelings were connected with it
that were attached to all the remains of more
ancient peoples. The treasure-hunter rarely
ventured on his search without having first
secured the aid of a magician for his protection
as well as for his guidance, for the same evil
spirits were believed still to haunt the ruins
underground, and it was hoped that by the
power of the conjuror they might not only be

rendered harmless, but be made to give information as to the exact spot where the treasure
lay. Numerous examples might be quoted of
such mediæval treasure-hunting on the Welsh
border, but it will be sufficient to give one
which appears to belong to the very site on
which we are now seeking treasures of another
description. An old manuscript chronicle of
the monks of Worcester, which is printed in
Warton's Anglia Sacra, and has preserved
numerous notices of events which occurred on
this border, informs us that in the year 1287,
at a place by Wroxeter, (that is near the
village), called "Bilebury," the fiend was com-
pelled by a certain enchanter to appear to a
certain lad and shew him where lay buried
"urns, and a ship, and a house, with an immense
quantity of gold." We easily recognize in the
objects described by the false Geomagog,
though not the material, the numerous figures
in bronze which are from time to time found
on Roman sites ; and the urns and ship may
perhaps admit of as easy an explanation. The
treasure-digger had to encounter sometimes a
worse opponent even than the fiend himself.
Treasure-trove belonged to the feudal lord, and
it was a right which he was inclined to enforce
with the utmost severity ; and the unfortunate
individual who was caught in the act of
trespassing against it found his way immediately
into a feudal dungeon, from which escape was
not always easy or quick. The learned historian
of this county, Mr. Eyton, has met with a record

from which we learn that some individuals towards the close of the thirteenth century were thus caught "digging" for a treasure at Wroxeter, and that they were thrown into prison. On their examination or trial, however, it appeared that, though they had dug for a treasure, they had not found one, and on this plea they had the good fortune to be set at liberty. This process of treasure-hunting had an effect injurious to the object of our researches. The mediæval excavator cared very little about antiquities as monuments of the past, and when, in digging a hole into the ground, he came upon a pavement, he broke it up without any scruple. It is to this cause, perhaps, that we must ascribe, in many cases, the damaged state in which we find the floors of the Roman houses, even when they lie at a considerable depth.

I have thus endeavoured to explain the manner in which a Roman town like Uriconium was ruined; how its ruin remained several centuries untouched, while a depth of earth was accumulating on the floors; how at a later · period the ruins themselves began to be cleared away, and a new accumulation of earth was formed over the lower part of the walls which had been left, until these could no longer be traced on the surface, except by the appearance of the crops in long periods of dry weather. This double accumulation of the debris of buildings has often led people to form erroneous conclusions, and in the account of a former partial excavation at Wroxeter, published by

the Society of Antiquaries, the writer has fallen
upon the rather old notion that the Roman
town had been burnt twice,—that he saw the
layers of burnt materials from two successive
burnings.

The effects of all these causes may be seen
in the excavations at Wroxeter,— the floor sometimes perfect and sometimes broken up ; the
walls of the houses remaining to the height of
two or three feet or more, as they were left by
the mediæval builders, when they carried away
the upper part of these walls for material ; the
original level of the Roman town on which its
inhabitants trod, strewed with roof-tiles and
slates and other material which had fallen in
during the conflagration under which the town
sank into ruin, and the upper part of the soil
mixed up with fragments of plaster and cement,
bricks and mortar, which had been scattered
about when the walls were broken up.

The site of Uriconium presents one great
advantage to the antiquarian explorer, that only
a small and not very important portion of the
area has been exposed to the most destructive
of all encroachments on its sanctity, modern
buildings ; while the situation and nature of
the ground has not required the deep draining
which would have cut through the ancient
floors, and these lie too far beneath the surface
to be touched by the plough.   It will be easily
understood that the preservation of such remains
depends much on the depth of soil which
covers them.   The Rev. T. F. More has dis-

covered and made considerable excavations in
a very extensive and most interesting Roman
villa, which occupies part of his beautiful park
at Linley Hall, near Bishop's Castle, but there
the position of the site, and perhaps other
circumstances, have caused the earth to accumu-
late much less rapidly, and the floors lay so
near to the surface that they have all been
destroyed. Where a fragment of the concrete
of the floor remained, it was hardly six inches
under the ground.

Our means of observation have hitherto ·
been so imperfect, that we can only form vague
conjectures as to the internal aspect and distribution of the buildings of a Roman town in
Britain. At the close of the Roman period the
towns were usually, if not always, surrounded
with defensive walls; but there are several
reasons for believing that the Roman towns in
this island were not walled until a comparatively late date, perhaps not till the domestic
dissensions and foreign invasions of the fourth
century. These town walls, when closely
examined into, are often found to contain
materials taken from older buildings of another
kind, which older materials themselves present
the debased style of architecture which belonged
to the declining age of the Roman power. The
long straggling line of wall which surrounded
Uriconium, as we may conclude from its very
irregularity, can only have been built at a late
date, after the city had gone on for ages increas-
ing in its extent. We are naturally led to

suppose that the public buildings would occupy
the central, or at least the more elevated part
of the town, and this has in several instances
proved to be the case. The discoveries made
by Sir Christopher Wren, seem to leave no
doubt that a Roman temple occupied the site
of the mordern cathedral of St. Paul's, in
London. But buildings of all sorts would seem
to have been mixed very confusedly together;
for we believe that in London, more recent
excavations have brought to light remains of
potters' kilns in close proximity to this temple.
In one or two instances, as at Aldborough, in
Yorkshire, (the Roman Isurium), and in some
of the small towns on the line of Hadrian's
Wall, in Northumberland, masses of the small
houses have been uncovered, and their appearance leads us to believe that the houses of a
Roman town in Britian were grouped thickly
together, that they were mostly separated by
narrow alleys, and that there were in general
few streets of any magnitude.

WE will now return to the spot where
the visitor has halted in view of the imposing
mass of Roman masonry, called the Old Wall,
situated, as has been stated, in a large tri-
angular field formed by the divergence of the
two roads. The Old Wall stands not quite
east and west, but sufficiently near it to allow
us, for sake of convenience, to call it east and
west. Its northern side is evidently the out-

side of a building, while there could be no
doubt that the southern side, on which the
springings of transverse walls and vaulted
ceilings are visible, was the interior. The excavations were begun on the February 3rd, 1859,
on the northern side, or outside, of this wall,
partly with the object of ascertaining the
depth at which the floors and the foundations
of the buildings lay under the present surface
of the ground, which, as we have said before,
was an important fact to ascertain. The bottom of the Old Wall was found at a depth
of fourteen feet, the last ten feet of which
was sunk in the natural substratum of sand,
so that the walls of the buildings in this spot
must have had originally very deep foundations.
It was found that this wall was continued
underground to the west, and excavations di-
rected towards the north brought to light
successively three walls running parallel, or
nearly parallel, to this first wall, the first of
these parallel walls being at a uniform distance
of fourteen feet from the Old Wall, the next
at a distance, also uniform, of thirty feet from
this wall, and the third at a distance from
the second of fourteen feet at the western
and sixteen at the eastern end, so that, as
the transverse wall at the eastern end of these
walls was not quite at right angles to them,
this large building was a little out of square.
This building, therefore, consisted of three
divisions, of which the central enclosure was
226 feet long by 30 feet wide, and appears
to have been paved in its whole extent with

small bricks, three inches long by one inch broad, set in zig-zags, or, as it is more technically called, herring-bone fashion. This description of pavement appears generally to have been used in passages and in open courts, and it seems probable, even from the magnitude of this enclosure, that it was not roofed. Nothing was discovered in it to throw any light on the object of so extensive a paved enclosure, but there could be little doubt that it must have been a public building of some importance. Portions of the capitals, bases, and shafts of columns were found scattered about in different parts of the area, which show that it was not wanting in architectural decoration, and on one of the pieces of wall-stucco, picked up in this part of the excavations, were three letters of what had been an inscription in large characters. Among other objects found here were a fragment of a very strong iron chain, the head of an axe, and an iron implement which appears to have been a trident, and to have been originally placed on a staff, perhaps an ensign of office. The appearance of the face of the Old Wall, which formed part of one side of the long narrow enclosure on the south of this central apartment, would lead us to suppose that this was an open alley, and this is confirmed by the other circumstances connected with it. In the continuation of the Old Wall to the westward, the lower parts of two doorways were found, which were approached from this alley each

by a step formed of a single squared stone,
which, therefore, may have been supposed to
have led from an exterior into an interior.
The corresponding long passage to the north
of the central apartment presented characteristics of another kind. At the eastern end
were found pavements of rather fine mosaic,
of which specimens and admirable drawings,
by Mr. George Maw, of Broseley, are preserved
in the Museum. Mosaic of this description
was not made to be exposed to the air, and
the building here must not only have been
roofed, but we have reason to suppose that
there must have been a room or rooms of
a character on which elegant ornamentation
would be bestowed.

The walls of this building, as we find them
underground, present from time to time dis-
continuations, or breaches, caused no doubt by
the breaking up of the walls for materials by
the mediæval builders, who sometimes went
deeper for them than usual; and it is very
likely that this may have been caused, in some
instances at least, by the circumstances that on
the site of these breaches were doors or passages, the jambs and ornamental parts of which
were formed of large stones which were more
tempting to the old excavators. With the
exception of these breaches, there are no traces
of doorways from one apartment of this building to the other. About the middle of the
northernmost wall there is a very wide breach
of this kind, which perhaps represents a grand
entrance from the north. Moreover, in carry-

ing the excavations further towards the north,
it was found that this northernmost wall of
the building formed the side of a street, which
was paved in the middle with round stones,
not much unlike the pavements of some of the
streets in Shrewsbury and other old towns as
they remain at the present day. The northern
wall just alluded to was traced eastwardly
until the edge of the field in which the excavations were carried on prevented the workmen
from going any further. Immediately to the
east of the building we have been describing
was a not quite rectangular enclosure, which,
from the appearance of the walls, was probably
a court-yard. A door-way, approached by a
stone step within the great inclosure to the
west, led into it. Beyond this, to the eastward,
was a much larger enclosure, which as far as it
was explored, had no tracings of walls or pavement within, and may possibly have been a
garden. At the western end of the great building, about the middle of the extremity of the
great central inclosure, indications were discovered which probably belonged also to an
entrance. These indications consisted of two
original openings in the wall, within which
were found, evidently in their original position,
in one a large squared stone, and in the other
two similarly squared stones placed one upon
another. One of these was bevelled off at the
outer edge into a plain moulding, and their
general appearance led to the belief that they
had formed the basis of something—perhaps

of large columns. Here, therefore, may perhaps
have been the principal entrance into the long
and extensive area* which occupied the middle
of this building. It faced the modern Watling
Street Road, which evidently represents another
street; and it thus seems to admit of no doubt
that this building formed the corner of two
principal streets of the Roman city of Uriconium.

We will now return to the long alley, as
we have ventured to call it, on the southern
side of the building we have been describing.
It has been already stated that there were
found in this alley two steps, formed each of a
large squared stone, attached to two doorways
in the western continuation of the Old Wall.
The more western of these two steps was very
much worn by the feet of the people who had
passed over it, as though it had led to some
place of public resort. It was at the more
easterly of these doorways that the excavations
were carried to the southward of the Old Wall.
This doorway apparently led into some open
court which communicated with domestic apartments. A trench carried directly southward
from the doorway, brought the excavators to
the semicircular end of a hypocaust, which had
warmed a considerable room thirty-seven feet
long by twenty-five feet wide, and which was in a
state of very perfect preservation when opened,
although the floor which once covered it had
entirely disappeared. The pillars, which were
formed of Roman square bricks, placed one
upon another without mortar, and of which 120

were counted, were above three feet high. This room has now been competely laid open, and on the western side has a complicated arrangement of walls, which evidently served some purpose connected with the heating of the hypocausts. A considerable quantity of unburnt coal was found here. The northern end of this hypocaust, the wall of which remained to the height of several feet, presents an imposing mass of masonry, and we learn from it the interesting fact that the Roman houses were plastered and painted externally as well as internally. The exterior of the semicircular wall at the north end of this hypocaust was painted red, with stripes of yellow. Near it lay an immense stone, hewn into the shape to fit the semicircular wall of the hypocaust, which had evidently formed part of a massive band of such stones at some height in the wall. A strong piece of iron is soldered into it with lead, for the purpose of attaching something to the building externally. A little alley, considerably wider than the spaces between the pillars of bricks, ran across this hypocaust, and through an opening in the wall, into another hypocaust, which was entered from without by a large archway, and this again was approached by a flight of three steps, each step composed of one large well-squared stone, descending from a square platform, which was apparently on a level with the original floors of the rooms. When the steps were uncovered, a broken shaft of a large column was found lying across them.

The platform at the bottom of the steps, or at
least the corner of it farthest from the arched
entrance to the hypocaust, seems to have been
used by the last occupiers of this building as
a receptacle for the dust swept from floors and
passages, for the earth, for about a foot deep on
the floor, was literally filled with coins, hair-pins,
fibulae, broken pottery and glass, bones of birds
and animals which had been eaten, and a
variety of other such objects.

To the east of the entrance to the hypocausts, a small room only eight feet square was
found, which had a herring-bone pavement like
that of the great inclosure to the north of the
Old Wall. A rather wide passage through the
eastern wall of this small room led into another
room with a hypocaust, the floor of which is
also gone. The pillars of this hypocaust were
rather more neatly constructed, but they seem
to have been considerably lower than those of
the hypocausts previously opened. This hypocaust was the scene of a very interesting
discovery. Abundant traces of burning in all
parts of the site leave no doubt that the city
of Uriconium was plundered, and afterwards
burnt by some of the barbarian invaders of
Roman Britain at the close of the Romano-
British period, that is, towards the middle of
the fifth century. The human remains which
have been met with in different parts, bear
testimony to a frightful massacre of the inhabitants. It would seem that a number of
persons had been pursued to the buildings

immediately to the south of the line of the
Old Wall, and slaughtered there; for in trench-
ing across what were perhaps open courts to
the south and south-east of the door through
the continuation of the Old Wall, remains of
at least four or five skeletons were found, and
in what appears to have been a corner of a
yard outside the semicircular end of the hypocaust first discovered, lay the skull and some
of the bones of a very young child. In the last
of the hypocausts we have been describing, three
skeletons were found, that of a person who
appears to have died in a crouching position
in one of the corners, and two others stretched
on the ground by the side of the wall. An
examination of the skull of the person in the
corner leaves no room for doubting that he
was a very old man. One at least of the others
was a female. Near the old man lay a little
heap of Roman coins, in such a manner as to
show that they must have been contained in a
confined receptacle, and a number of small iron
nails scattered among them, with traces of
decomposed wood, prove that this was a little
box, or coffer. The remains of the wood are
still attached to two or three of the coins. We
are justified from all these circumstances in
concluding that in the midst of the massacre of
Roman Uriconium, these three persons—perhaps
an old man and two terrified women—had
sought to conceal themselves by creeping into
the hypocaust; and perhaps they were suffocated
there, or, when the house was delivered to the

flames, the falling rubbish may have blocked up the outlet so as to make it impossible for them to escape. It is not likely that they would have been followed into such a place as this hypocaust. These coins were 132 in number, and the following description of them has been given by Mr. C. Roach Smith :—

TETRICUS. One much worn, of the *Fides Militum* type..   1

CLAUDIUS. One, *rev.* CONSECRATIO ; an eagle   1

Constantine the Elder. *Obv.* CONSTANTINVS. MAX. AVG. Head diademed, or wreathed, to the right. *Rev.* GLO-RIA EXERCITVS. Two soldiers with spears and shields, standing; between them two standards ; or (in three instances) a single standard.

    Mint Marks (exergual letters): P . CONST., 3 ; TR . P.. . 6 ; S . L . C. 1 ; illegible, 3 ; total   13

CONSTANS. *Obv.* Much worn or decayed. *Rev.* FEL . TEMP . REPARATIO. The emperor holding a globe and a standard, standing in a galley rowed by a Victory. This coin is altogether much worn. It possibly may have been plated   1

CONSTANTINE II. *Obv.* CONSTANTINVS . IVN . NOB . C. Laureated head, to the right; bust in armour. *Rev.* GLORIA EXERCITVS. Two soldiers standing; between them two standards, and on the same a wreath, or other object, in the field.

    Exergual letters : TR . P. or TR . S.. 15 ; P . L . C.. 9 ; CONST., 3 ; illegible, 9 ; total   36

CONSTANTIUS II. *Obv.* T . L . IVL . CONSTANTIVS . NOB . C. Laureated head, to the right; bust in armour. *Rev.* GLORIA EXERCITVS. Two soldiers, &c., as on the coins of the preceding.

    Exergual letters : TR . S., 3 ; P., 1 ; SMTS, 1 ; total   5

JULIAN. A plated denarius. *Obv.* FL . CL . IVLIANVS :
P . F . AVG. Diademed head to the right. *Rev.* VOTIS
V MULTT . XX., within a wreath 1

HELENA. *Obv.* T : L . IVL . HELENAE AVG. Head to the
right. *Rev.* PAX PVBLICA. A female figure standing
and holding in the right hand a branch, and in the
left hand a *hasta pura.* In the field, a cross ; in the
exergue, TR . P. Another without the cross. Total 2

THEODORA. *Obv.* FL . THEODORAE. AVG. Head to the
right. *Rev.* PIETAS ROMANA. A female standing suck-
ling an infant: in the exergue, TR . P. 1

URBS ROMA. *Obv.* VRBS ROMA. Galeated head of Rome,
to the left. *Rev.* Romulus and Remus nursed by the
wolf ; above, two stars : on two, two stars and a wreath.
In the exergue: PL . C., 11 ; TR . P . or TR . S., 10 ;
illegible, 3 ; total........................... 24

CONSTANTINOPOLIS. *Obv.* CONSTANTINOPOLIS. Bust of per-
sonified Constantinople, helmed, and holding a sceptre,
to the left. *Rev.* A winged Victory, with *hasta pura* and
shield ; her feet upon the prow of the galley, to the left.
Exergual letters: TR . P., 20 ; P . L . C. or S . L . C.,
9 ; O . SIS. 1 ; S . CONST., 1 ; illegible, 3 ; total.. 34

VALENS. *Obv.* D . N . VALENS.... Diademed head, to
the right. *Rev.* SECVRITAS.... Victory with wreath
and palm branch marching to the left. Much corroded 1

Rude copies of some of the foregoing 6

Extremely corroded........................... 6

Total number 132

D

This is, I believe, the first instance which has occurred in this country, in which we have had the opportunity of ascertaining what particular coins, as being then in daily circulation, an inhabitant of a Roman town in Britain, at the moment when the Roman domination in this country was expiring, carried about with him. Mr. Roach Smith, speaking of the great majority of these coins, these of the Constantine family, remarks to me—" I suspect these coins were sent into Britain even after the time of Valens, because they all are comparatively sharp and fresh. It is not improbable that the procurators at Treves and at Lugdunum may have had large stores of these coins by them, which they sent out at intervals." A consideration of these coins gives us an approximation, at least, towards the date at which Uriconium must have been destroyed; Mr. Roach Smith agrees in the opinion that a comparison of them points to the very latest period previous to the establishment of the Anglo-Saxons. At a later period the freshly struck coins of the Constantine family could not have been brought over. They show us that at that time the great mass of the circulating medium consisted of coins of the Constantine family, which again explains to us why the first coinage of the Anglo-Saxons was nearly all copied from the coins of the emperors of that family. Again, the care with which these small copper coins (for only one is of plated silver) seem to have been hoarded up, and the anxiety of their possessors to preserve them in the midst of a frightful calamity, may perhaps assist us in forming an estimate of the relative value of money at this period.

The rooms which joined up to the south side of the Old Wall, and which have been more recently uncovered, were five in number, and it appears from the remains, which are distinctly visible on the face of the Old Wall, that they had vaulted roofs of the kind technically called barrel roofs. In one of these rooms was found a quantity of burnt wheat, which would lead us to suppose that this might have been a store room. The most easterly of these rooms has had the interior surface of its walls ornamented with tessellated work instead of fresco-painting; the lower edge of which, consisting of a guilloche border, still remains. The floor below has a plain pavement of small white tessellæ, and .is apparently that of a bath. To the south of these rooms a long passage was discovered, which appears to have communicated at one end with the floor of the room in the hypocaust of which the skeletons were found. In this passage was a square pit of very good masonry, through which a drain runs, nearly north and south. The stucco of the southern face of the wall, forming the southern side of the passage just alluded to, presented an inscription scrawled in large straggling characters incised with some sharp pointed instrument, and closely resembling in character similar inscriptions which have been found on walls in Pompeii. When first uncovered, two lines of this inscription, perhaps the whole of it, seemed to have been perfectly well preserved, but before anybody had had the opportunity of examining it, two casual visitors with walking sticks, amused themselves with breaking off the plaster, in order apparently to try its strength, and were not observed by the workmen until the first line had been competely destroyed, and the second, which had been a shorter one, was very much broken into, though just enough remained to show that it must have been written in Latin. Even this small remnant was nearly destroyed during the interruption of the excavations, and not a trace of it can now be seen. Thus all the advantages of a discovery which might have been singularly important for our knowledge of the state of Britain at this period, have been lost through mischievous wantonness.

During the month of May, 1859, the work of the excavators was interrupted; when it was resumed, they proceeded to explore the building to which these hypocausts belonged, beginning from the side of the field adjoining to the Watling Street Road,—that is, from the side of one of the main streets of the old Roman town,—and they found walls in the line, or nearly in the line, of the western wall of the great public building just described. Another street has since been discovered to the south, running east and west, parallel to that met

with to the north of the buildings first excavated.
The excavations have since that time been
followed in various parts of the two acres first
inclosed by the Excavation Committee, and a
large extent of ruins is now laid open. But I
will here interrupt my narrative, while I give
an account of the general character of the
buildings, the ruins of which have already been
brought to light.

As yet the excavations on the site of
Uriconium have not been carried far enough
to enable us to form any idea of the general
distribution of the Roman town, but it is evident that the building on which the excavators
are employed were inclosed by three main
streets, crossing at right angles, forming a
square mass. It has been stated that the few
discoveries hitherto made as to the character
of the streets in the Roman towns in Britain
would lead us to think that they were little
more than narrow alleys, but this was certainly not the case with these three streets of
Roman Uriconium, which seem to have been
fine wide streets, and in the one to the north,
the pavement of small round stones appears to
have occupied only the middle part of the
street, designed probably for carriages and

horses. A tolerably wide space on each side seems, as far as can be traced, to have been unpaved. But, although we have as yet made little advance towards discovering the general character of Uriconium as a city, and the manner in which the houses were distributed over the Roman town, we had found sufficient fragments of different kinds to give us a tolerable notion of the houses themselves.

The average thickness of the walls of a house even where they only separated one small room from another, was three feet. They are rarely less than this, and it is only in one or two cases of what appeared to be very important walls that they exceed it, when they reach the thickness of four feet. This measure of three feet was no doubt a well understood one for the wall of a house, and it was continued in the middle ages, when, in ordinary dwellings, only the division walls between house and house were of solid masonry. Municipal regulations then fixed these partition walls at a minimum of three feet in thickness, the cause of which limitation was probably the fear of fires ; and in these mediæval municipal regulations, it was further ordered, that closets or cupboards in the wall should in no case be made more than one foot deep, so that if your own cupboard and your neighbour's happened to back each other, there would still be a foot of solid masonry between the two houses. And the masonry of the Romans may well be called solid. Its character may

be seen perhaps to most advantage in the Old
Wall above ground. The process of building
seems to have been to raise first, gradually,
the facings of neatly-squared stones, supported,
no doubt between frames of woodwork, the
supports of which left holes which are still
seen in the face of the wall. The interior was
then filled up with rubble mixed with liquid
and apparently hot cement, which formed the
mass of the wall, and in setting has become
in course of time harder than the stones
themselves. After a certain number of rows
of facing-stones, the Roman builders almost
invariably placed a string-course of broad thin
bricks, the object of which is not at all evident, for they do not go through the wall so
as to form real bonding-courses. The Old
Wall still standing in probably nearly its
original height, will also give us a notion of
the elevation of the principal houses of the
Roman towns.

In spite, however, of this rather considerable elevation, which, reckoning for dilapidation
at the top and the portion buried under ground,
cannot have been much less than thirty feet,
it seems nearly certain that the Roman houses
in Britain had no upper stories, and that all
the rooms were on the ground floor. No
traces of a staircase have ever been found,
and all the fragments which are met with,
indicate that the rooms were open to the roof.
These roofs appear to have been of substantial
construction, and were probably supported on

a strong frame of woodwork. The common coverings of the Roman houses of this island consisted of large square tiles with strongly flanged edges, and these tiles being joined side to side, a curved tile forming the half of a cylinder was placed over the flanges of the two tiles which joined, thus holding them together, and at the same time protecting the juncture so that rain could not pass through it. These tiles, and the manner in which they are arranged, will be understood by our figures, (pl. IV., figs. 1. 2. 3.) The Roman houses were also very commonly roofed with slates, or rather flags, and this appears to have been the more usual description of roofing at Uriconium. These roof-flags are found scattered about abundantly on the floors, sometimes unbroken. They are formed of a micaceous laminated sandstone, which is found on the edge of the north Staffordshire and Shropshire coalfield, at no great distance from Wroxeter, and must have produced a glittering appearance in the sunshine. Their form is represented in our cut, (pl. IV., fig. 5); it was that of an elongated hexagon, with a hole at one end, through which an iron nail was passed to fix it to the wooden frame-work. The nail is often found still remaining in the hole. These flags, which are very thick and heavy, were placed to lap over each other, and thus formed a roof in lozenges or diamonds, as represented in fig. 6. Slates, forming one half of the hexagon (fig. 4), were placed at the top of

the roof, so as to make a strictly horizontal
line. It is a curious circumstance that in the
illuminations of Anglo-Saxon manuscripts we
find roofs of houses which evidently represent
both these methods, and which appear, therefore, to have been continued long after the
Roman period. In fact they are still used
in Yorkshire, and perhaps in other counties,
and have been used very recently on the Welsh
border. In the towns which were the head-
quarters of a legion, as at Caerleon, Chester,
and York, or which had been occupied for
some length of time by legionary detachments,
we often find the name and number of the
legion stamped on the roof-tiles. These roof-
tiles were frequently used for other purposes.
They are sometimes employed in the string-
courses in walls, when the builders appear to
have run short of the ordinary square tiles
or flat bricks; and they are still more frequently used to form the beds of drains and
aqueducts, when the flanged edges were turned
up, and set in the cement, formed the side of
the water-course. A very good example of
this use of the roof-tiles may be seen in the
drain at Wroxeter mentioned above.

Internally, the walls of the Roman houses
were covered with fine hard cement, which was
painted in fresco, that is, the colours were laid
on the cement while it was wet, and they thus
set with it, and became almost imperishable. In
some of the houses in Roman Britain, and especially in the large villas, the internal walls were

covered with fine historical subjects as in the walls of Pompeii, and sufficient remains have been found in this island to show that they were here also executed in no mean style of art. Nothing of this kind has yet been discovered in Uriconium; but numerous fragments are picked up in the diggings, on which the colouring is perfectly fresh, and which exhibit portions of designs which are always elegant and in good taste. In one case a piece of the stucco from the internal surface of a wall contained some letters of an inscription. One of the walls near the hypocaust where the three skeletons were found presented a singular and rather laborious method of ornamenting its interior surface. Instead of being painted, it was tessellated, the surface being covered with tessellæ, one half of an inch by three-fifths in dimension, set in the cement, alternately of dark and light colours, in horizontal lines, so as to produce somewhat the appearance of chequer-work. Perhaps when entire, it presented an ornamental pattern. I have already stated that a similarly tessellated wall was found in the easternmost part of this line of rooms. Circumstances have come to light which show that the exterior of the walls of houses were also plastered and painted. The exterior of the semicircular end of the largest hypocaust yet opened was thus plastered over, and painted red with stripes of yellow.

It is worthy of remark that in the walls, to the certainly not very great elevation they now generally reach, few doorways are discovered, a circumstance which is by no means easily explained. Small rooms are found without any apparent means of access. Perhaps, in such cases, the doorway was at a certain elevation in the wall, and was approached on both sides by wooden steps, which have long perished, and left no traces of the means of entrance. Of course, none of the walls of the houses remain sufficiently high to enable us to judge of the manner in which light was admitted into the rooms, whether from side windows, or from openings in the roof. Probability, however, is in favour of roof-windows being in common use, and an interesting circumstance connected with the excavations at Wroxeter seems decisive as to the material of the windows. Considerable quantities of fine window glass have been found scattered over the floors of the houses, of an average thickness of full one-eighth of an inch, which have been duly deposited in the Museum at Shrewsbury. It is the more curious as it has been the common opinion, until recently, that the Romans, especially in this distant province, did not use window-glass; and the fragments of window-glass which have been found more recently in the excavations on the sites of Roman villas have been much thinner than that found at Wroxeter, and of very inferior quality. It is evident, that some of the rooms, all the walls of which were only walls of separation from other rooms, must have received light from above, or have been quite dark.

I must now describe a peculiar characteristic of the domestic economy of a Roman house in Britain, and in the other western and northern provinces of the empire. The Romans did not warm their apartments by fire lighted in them, as was the case in the middle ages, and in modern times, but by hot air circulated in the walls. The floor of the house, formed of a considerable thickness of cement, was laid upon a number of short pillars, formed usually of square Roman tiles placed one upon another, and from two to three feet high. Those of the largest of the hypocausts yet found at Wroxeter were rather more than three feet high. Sometimes these supports were of stone, and in one or two cases in discoveries made in this country, they were round. They were placed near to each other, and in rows, and upon them were lain first large tiles, and over these a thick mass of cement, which formed the floor, and upon the surface of which the tessellated pavements were set. Sometimes small parallel walls, forming flues instead of rows of columns, supported the floors, of which an example has already been found in the excavations at Wroxeter. Flue-tiles,—that is, square tubes made of baked clay, with a hole on one side, or sometimes on two sides,—were placed against the walls end-ways, one upon another, so as to run up the wall. These arrangements,—which were called hypocausts, from two Greek words, signifying *heat underneath*, and were used in Italy and Greece chiefly for warming baths, are

represented in plate IV., *fig.* 7, where AA is the
floor of cement, B B the pillars supporting it,
and C C the flue-tiles running up the wall of
the room. They had an entrance from the
outside somewhat like the mouth of an oven,
and fires being lighted here, the hot air was
driven inward, and not only filled the space
under the floor, but entered the flue-tiles by the
holes in the sides, was carried by them up the
inside of the wall, and no doubt had some way
of escape at the roof. The ashes and soot of
the fires have been found in the hypocausts at
Uriconium, just as they were left when the city
was overthrown and ruined by the barbarians.
The ashes are chiefly those of wood, but considerable remains of mineral coal have been
discovered. These hypocausts must sometimes
have become clogged and out of order, and it
would be necessary to cleanse them, as people
in aftertimes cleansed chimneys. A sort of
alley across the middle of the large hypocaust
last-mentioned, was probably intended for this
purpose. It communicated with another hypocaust adjoining it to the north by a doorway,
and this other hypocaust was entered by a
rather large archway at the foot of the steps
already mentioned. People appear to have
been sometimes satisfied with having the hot
air merely under the floor, and the flue-tiles
were not always used. Comparatively few of
them, indeed, have been yet found in the
hypocausts of Uriconium.

THE requirements of argiculture have rentered it necessary to cover up again all the excavations to the north of the Old Wall, and the walls of the great public building at the corner of the two streets can no longer be seen by the visitor. A piece of ground, however, immediately to the south of the Old Wall has been taken by the Excavation Committee at Shrewsbury upon a rent, and on this piece of ground the excavations are now carried on. It forms a parallelogram, 319 feet long, by 279 feet wide, containing an area of exactly two acres, including the Old Wall at its northern edge. This piece of ground has been strongly fenced round with hurdles, and it is entered by a gate from the Walting Street Road. By the liberality of the Excavation Committee the public are admitted to this inclosure freely, and it is to be hoped that the visitors will acknowledge this liberality by carefully abstaining from committing any injury on the Roman remains, or by walking upon or entering into the parts in the course of excavation.

The plan annexed (*p.* 5) of the excavations now in progress will enable me to explain them to the visitor. The darkly-shaded mass, *a a*, represents the Old Wall, or portion of Roman masonry, standing above ground ; to the north of which lay the extensive building formed by the walls *b b*, *c c*, *d d*, running parallel to the Old Wall. The wall *d d*, bordered upon a wide street. To the east of these walls lay an inclosure, *c*, perhaps a courtyard, and a large

space, *f,* which has been conjectured to have
been a garden, but which has been very imper-
fectly explored. All these remains have been
explained before ; they have been buried again,
and the ground is now covered with crops.
The Old Wall, which stands just within the
north-eastern corner of the space separated from
the rest of the field by a fence of hurdles, now
forms the northern boundary of the excavations.

The visitor is introduced into this space by
a gateway from the road, nearly at its north-
western corner. Opposite this gateway he will
see an apartment, which the excavators are
now in the course of exploring. It is nearly a
square, and is about thirty-four feet in its
longest dimensions. The side towards the
street seems to have been open, or at least the
masonry of the wall presents the appearance
of having had wide folding doors, or a frame-
work of wood of some kind in two compartments 6, 6. In the centre of the room is a large
pier of masonry (1), perhaps a table for workmen.
More towards the north-western corner, a sort
of furnace or forge (2) was found, built of red
clay, with a hole or cavity in the upper part
sufficiently large for a man to thrust his head
in. As the surface of the cavity, internally,
is competely vitrified, and as there was much
charcoal strewed about, there can be no doubt
that the cavity had been occupied by a very
fierce fire. A low wall has been traced, running
across the room east and west in a line with
this furnace ; and two transverse low walls of

similar character. Upon the low wall a little
behind the forge (at 3), the excavators came
upon what was supposed to be the lower part
of a column with its base; but it is formed
roughly, and I think it more probable that it
was a stone table for the use of the workmen at
the furnace. It was at first supposed that this
might belong to a colonnade running along
the wall; but no trace of such a colonnade has
been found, although a large piece of a shaft of
a column lies in the middle of the room. This
column, however, is of larger dimensions than
the supposed base (3). Had such a colonnade
existed, it seems so little in accordance with
the existence of a forge, that we might be led
to suspect that the room had, at some late
period, been diverted from its original purpose,
and occupied by a worker in metals, or even
in glass, as fine specimens of glass were found
scattered about, and also many fragments of
metal. But objects of all kinds seem to have
been thrown about in such a manner, when the
town was plundered, that it would be unsafe
to argue upon the purpose of any particular
building, merely from moveable articles found
in it. Among other things found in this room
were nearly a dozen hair-pins, two of which
were much more ornamental than any we had
found before; a much greater quantity of fragments of Samian ware, and of higher artistic
merit, than had previously been met with in
one spot; a portion of a large bronze fibula; a
number of coins, and other things. One of

the vessels of Samian ware is a fine bowl with figures in high relief, representing a stag-hunt. Upon the low wall of the sill (6) a number of copper Roman coins (about sixty) were found together; and near them the frag-ment of a small earthern vessel, in which probably they had been carried by some one who dropped them there as he was hurrying out of the place. Turning from the gate of the field to the right, or south, along the inside of the hedge, the visitor will come to a porton of uncovered wall, *h h*, running north and south, upwards of eighty feet, in which there are two entrance gateways, *i, p*. The first of these is about twelve feet wide, and was approached by a sort of inclined plane, formed of three large squared masses of stone, each about four feet square by eleven inches in thickness. The other entrance, which was only five feet wide, was approached by two steps, each similarly formed 'of one mass of stone; of which the lower step is worn very much at its south-west corner, in a manner to lead us to believe that the great majority of the people who passed through this entrance came up the street from the south. The upper step, or stone, is so much worn by the feet of those who passed over it, that it broke into three pieces under the work-men's picks. On one side of it there is a deep hollow, representing nearly the form of a small human foot, which seems to have been scooped into the stone for some purpose with which we are not acquainted. These two entrances lead

E

into one square court, the floor of which, proved
by the steps and inclined plane to have been
on a higher level than the street without, was
paved with small bricks laid in herring-bone
work, like the great inclosure to the north of
the Old Wall. It is found to have been much
damaged and mended in ancient times, which
seems to countenance the supposition that the
wide entrance and the inclined plane by which
it was approached were intended for horses
and perhaps for carts or for heavy barrows.
Among the objects found in excavating here
was a portion of a horse-shoe. On each side
of this court a row of chambers is found, *m m m*,
four on the north side and four on the south,
from ten to twelve feet square. The western-
most of these chambers, on the north side of
the court, has been cleared out, and was found to
be ten feet deep, with a low transverse wall at
the bottom, the object of which is at present
quite inexplicable. A quantity of charcoal
was found in this room, as though it had been
a store-room for that article. One of the other
rooms, on each side of the court, seemed to
have been a receptacle for bones, horns, &c.;
and as some of these had evidently been sawn
and cut, and others partly turned on a lathe,
they suggested the idea of having belonged to
manufacturers of the various objects made of
this material which are found so commonly in
the course of the excavations. They may,
therefore, have been the magazines of manufacturers and tradesmen, a notion which is

somewhat confirmed by the circumstance of
several weights of different sizes having been
found in this part of the excavations ; or they
may have been mere depôts for the stores and
refuse of a large mansion or other establishment.
These rooms are, perhaps, all deep like the one
already cleared out, but it is remarkable that, as
high as the walls remain, that is, about two
feet above the floor of the court, there is no
trace of entrances to them, which must, therefore,
have been rather high in the wall, and they were
entered perhaps by a ladder.

The back part of this court consists of a
long narrow inclosure, which is divided into
compartments by four transverse walls proceeding from the western wall about halfway across
the inclosure, thus leaving a passage along the
eastern side. These compartments have much
the appearance of small shops or stalls for
selling, and seem to confirm the notion that this
building may have been a market-place. The
workmen, finding a doorway in the wall of the
back of this inclosure, at n in the plan, a trench
was carried through the ground to the eastward.
At about twelve feet from the opening at n,
they came upon a wall at h, running parallel to
the wall o o of the court, and beyond this they
found first a narrow passage, and then a rise
with a pavement of cement which extended
some four or five feet, and then suddenly sank
to a floor of large flag-stones, at a depth of
upwards of four feet from the floor of cement.
This flagged floor, the position of which is

marked by the letter q in the plan, was perhaps
a reservoir of water; the bottom was found
covered with black earth filled with broken
pottery and other things, such as may easily
have been supposed to have been thrown into
a pond. The water appears to have been only
between two and three feet deep, as the floor
on the opposite side runs about level with the
ledge or step just mentioned, and is continued
eastward until, at r, we come upon the rather
massive walls of a building, the nature of which
cannot be determined without further investiga-
tion. At a short distance within this wall, at
a depth of about three feet below the cement
floor, we find a floor, at s, about ten feet wide
by thirty long, formed of flat Roman tiles,
twelve inches by eighteen inches square. This
floor has been uncovered, and as there was an
indentation in the middle which seems to
indicate that it was hollow underneath, a hole
was made there, but it lead to no discovery.
This seems also to have been a tank of water,
perhaps a cold water bath. The cement floor
was continued easterly until it was terminated
by a wall, t, which ran at right angles to the
eastern end of the Old Wall, and appears to be
the eastern termination of the buildings now
in course of exploration. The earth and rubbish from the excavations have been here
thrown into a great mound, from the top of
which the visitor can enjoy the bird's eye view
of the excavations. A few yards to the north,
he will come to the important line of excava-

tions nearer to the Old Wall. A small chamber,
about eight feet square, with a herring-bone
pavement in very good preservation, projects
beyond the line of this eastern wall at *u* in
our plan. To the west of this is a small
hypocaust *r*, the floor of which has been a
little lower than that of the room *u*. In this
hypocaust were found the remains of two
skeletons, one of which was that of a young
person. The northern wall of the room *v* is
particularly interesting, because in its whole
height of full nine feet, it presents the remains
of the lines of flue-tiles which ran up it, hardly
an inch apart, and which show that this room
must have been intended to be very much
heated. It was, perhaps, a *sudatorium*, or
sweating room. The opening from *u* to *v*
occupies nearly the whole width of the former
room, and was perhaps closed by a wooden
door. On the western side of the hypocaust,
at *w*, the wall has a sort of basement, formed
of large stones, scooped out in a singular manner,
the object of which is by no means evident.
We here come upon a series of passages *x*, to
the north of which were four rooms, *z z z z*,
extending to the Old Wall. On the face of
the Old Wall we can distinctly trace the
springing not only of the walls of division
the lower parts of which are found underground, but of the vaulting, from which it
appears that these rooms are technically called
the barrel-roofs of masonry. They were slightly
explored at the beginning of the excavations,

and in one of them was found a quantity of
burnt wheat, as though it had been a store-
room.

In the passages alluded to, there is at *y*,
a square pit, somewhat like what might be a
cesspool, of very good and substantial masonry,
at the bottom of which runs north and south
a very well formed drain, the bed of which is
formed of large roof-tiles. To the south of this
is a hypocaust, A, which differs from the other
hypocausts yet opened, in being partly formed
of low parallel walls instead of rows of pillars.
On the wall of the passage leading to this
hypocaust from the east was found the inscription mentioned at page 45. Westward from
the hypocaust A, but without any apparent
communication between them, was another
hypocaust, B, which had been constructed in
the usual manner, the floors supported by rows
of low columns, formed of square thin bricks.
It was in this hypocaust that the three skeletons
mentioned before (*p.* 41) were found, the man
who possessed the money crouching in the
north-west corner, and the two persons supposed to be women, extending along the side
of the northern wall. The opening into this
hypocaust was through its southern wall, from
the interior court, so that the fugitives must
have crept along the whole length of the
hypocaust to reach their place of concealment.
The part of this interior court, immediately
adjacent to this hypocaust, which has been
excavated to some extent, presents several

interesting features. A breach in the eastern
boundary wall had been newly repaired with
much inferior masonry at the time when the
city of Uriconium was taken and destroyed ;
and it is a curious circumstance that some
large pieces of stone lie here on the floor of
the court, unfinished by the masons, as though
repairs and alterations in the buildings were
going on at the very moment of the final
catastrophe. Adjoining to this hypocaust, at
its north-west corner, is a square room c, with
the herring-bone pavement, exactly like that
at *u* in character and dimensions which had
opened into the room above the hypocaust B,
much in the same manner as *u* opened in the
room *v*. Separated from this room by a wall,
but apparently without any communication
with it, is an interesting staircase, D, leading
down to the entrance to a larger and apparently
more important series of hypocausts. This
staircase descended from a square room, about
the same size as the room c, which had a
smooth pavement of cement. It is composed
of three steps, each formed of a large squared
stone. A part of the space at the bottom,
the north-eastern corner, appears to have been
used by the later Roman inhabitants of this
building as a receptacle for the sweeping
of the floors, and when it was first opened the
earth, to the height of about sixteen or eighteen
inches from the floor, was filled with all kinds
of objects, such as coins, hair-pins, fibulæ, needles
in bone, nails, various articles in iron, bronze,

and lead, glass, broken pottery, bones of edible
animals and birds, stags' horns, tusks and hoofs
of wild boars, oyster shells, in one of which
lay the shell of a large nut, &c.   A large shaft
of a column lay across the steps.   The Roman
masonry here is very good.   To the right hand,
towards the south, a rather large arch, turned
in Roman bricks, led into the hypocaust E, a
doorway in the southern wall of which formed
the communication between this hypocaust and
the still larger hypocaust F.   The latter had
supported what must have been a handsome
room, which was about fifty feet long, including
the semicircular northern end, by thirty-five
in breadth.   When first opened, this hypocaust
was in a state of preservation in which such
buildings are seldom found in this country.
A hundred and twenty columns of bricks were
counted, most of them at their original height,
of rather more than three feet.   At the north-
eastern corner, the columns supported a small
portion of the floor in its original position.
It is a mass of cement, eight inches thick,
with the upper surface, which no doubt had
formed the floor, perfectly smooth.   During
the time that the Excavation Committee were
excluded from the field, all the pillars of this
interesting hypocaust were thrown to the
ground, and a great part of the bricks which
formed the supporting columns were broken
to pieces—even the pieces of the floor and its
supports at the north-east corner were overthrown.   A very exact drawing of the latter,

however, had been preserved, which served as
a pattern for restoring it; and it is to the
ingenuity and labour of Dr. Henry Johnson
that the public owes the restoration of this
hypocaust as far as it is possible to restore
it.

Returning to the steps by which these
hypocausts were entered, at D, the floor from
which we descended appears to have an
opening of some kind to the west, which
looked down upon a court outside the semicircular end of the hypocaust F, which from
this point presents to the view an imposing
mass of masonry. In the corner just under
this opening the remains of a very young
child were found, which we might almost
imagine to have been slaughtered in the room
above, and thrown out into the court. This
court, or open space, seems to have been continued to the wall a a, and to have been
entered by a doorway in that wall at g, which
was approached from the passage to the north
by a step formed by a large squared stone.
On the outside of the semicircular end of the
hypocaust F, lay, as if it had fallen or been
thrown down, an immense stone, carefully
worked into the shape of the arc of a circle,
and no doubt forming one of a course at some
unknown elevation in the wall. On the out-
ward side of it, a large iron pin was soldered
into it with lead, evidently for the purpose of
attaching some weighty object on the outer
side of the building.

Another step and doorway in the wall *a a* was found at *h*, which must have been much more frequented than the other, for the stone which formed the step was worn in an extraordinary degree by the rubbing of footsteps. It led to an inclosure, P, which presents the appearance of having formed public *latrinæ*; and which is separated by a long narrow inclosure from the room already described, as apparently the shop of a worker in metals.

Such is a brief and general description of the ruins of Uriconium, at present open to the visitor. The real character of the buildings we have been describing appeared for a while very doubtful. The first discoveries led to the belief that it was a great mansion, perhaps the principal mansion in the Roman city, the residence of the chief municipal officer; but in this case we might have expected to find some very fine Mosaic or tessellated pavement, specimens of which had been met with in other parts of the area of the town. On the contrary, all the floors yet discovered to the south of the Old Wall, with the exception of those of herring-bone brickword, and that of a supposed bath, seemed to have been of mere smoothed cement. This led us to suppose that we were still exploring buildings erected for some public purpose. A comparison of the character of these various buildings leaves no room for doubting that they belonged to the public baths of Uriconium; and further excavations to the south and west showed that

they formed an extensive square, (*k,k,k,k,*) the
northern side of which was formed by the
Old Wall and its continuation westward; and
the southern side of which bordered upon the
other street running east and west, the pavement of which, similar to that of the street
at *l*, has been uncovered in its whole extent
along the line L L. The western and southern
sides of the square were formed by a wide
gallery or cloister (*k,k,k,*) no doubt the ambu-
latory, which was considered as an important
part of the public baths of the Romans. The
ground to the eastward, in which no buildings
could be traced, may have been gardens which
were also usually attached to the baths of the
Romans.

Having once decided that the building
we have thus explored is the public baths,
another equally interesting question arises out
of it. The public baths of the Roman towns
in Britain, are not unfrequently mentioned in
inscriptions commemorating the repairing or
rebuilding of them; but it is a circumstance
of some importance that this building is combined with the basilica, or town hall. Both
seem to have participated in the same accidents,
and to have undergone decay together. Thus
an inscription found at Lanchester, in Cumberland (supposed to be the Roman town of
Epiacum) speaks of the baths and basilica
(BALNEVM CVM BASILICA); and at Ribchester,
in Lancashire, the baths and basilica (BALINEVM
ET BASILICAM) were rebuilt after having fallen

into ruin through age. We are, therefore, I
think, justified in concluding that the two
great public buildings, the baths and the
basilica, usually joined each other; and I
think we may venture further to assume that
the large building to the north of the Old
Wall, the remains of which are now covered up,
was the basilica of Uriconium. The propor-
tions of this building are rather extraordinary,
and cannot be easily explained; but it is
probable that in a provincial town the basilica
served a variety of purposes. An inscription
found at Netherby, in Cumberland, speaks of
a basilica for practice in riding (BASILICAM
EQVESTREM EXERCITATORIAM.)

We may now proceed a little further in
identifying the topography of the ancient town.
The line of the buildings we have traced parallel
to the Watling Street road is at some distance
within the edge of the field; and I believe
that, when the farm buildings were erected on
the opposite side of the road, what appeared
to be the front of buildings facing the opposite
direction, were found likewise at some distance
within the field. This, with the road, would
make a very wide space; very much wider
than either of the two transverse streets.
Moreover, a glance at the plan will shew
that, beyond the transverse street to the south,
this wide space became considerably narrowed;
and in fact it seems to have been reduced to
the width of an ordinary street. It is my
belief that this wide space was the forum

of Uriconium; and in that case it is rather remarkable that the basilica held here exactly the same place, in regard to the forum, as at Pompeii.

We have thus already brought to light a very interesting portion of the ancient Roman town, and have learnt something more than we knew before of the character and economy of the Roman towns in Britain. The basilica, as we have seen, came up to the front of the street, and formed the side of a transverse street; but this was not the case with the baths, for a space of some width between them and the forum was occupied by other buildings, which I have already described.

Other apartments surrounding the metal-worker's shop are in the course of exploration, and will, I think, make us better acquainted with the character of the whole of this line of buildings which looked upon the open space which I have supposed to be the forum. I have already said that this open space contracts to the south of the transverse street L L, in what has been no more than the breadth of an ordinary street, which ran down towards the river. A gutter, very well made, of carefully squared stones, and remarkably well preserved, runs near the houses on the eastern side of the street; the only side which at present can be explored, as it is near the edge of the Watling Street Road. It runs very near the walls of the houses, is a foot wide, and about a foot deep, and from place to place square

stones are laid in lozenge-fashion, apparently
intended for stepping stones, but they must have
stopped the current of water down the channel.
The buildings at this corner consist of small
rooms, and were probably private houses. The
existence of walls running parallel and transverse to the street L L, has been ascertained
along the whole length of its southern side;
but they have not yet been sufficiently explored
even to be laid down in the plan.

———

THE objects of antiquity found in the course
of the excavations have been so often alluded
to, that the visitor will no doubt expect at
least a brief and general description of them.
I have already described those which illustrate
the building and construction of a house,
and we naturally continue the description by
turning to those articles which belong especially
to domestic life. Of this class, the most numerous
division, and that which strikes us first, is the
pottery,—of which certainly the most remarkable to the general observer is the ware
resembling in colour and general appearance
bright red sealing wax, known commonly as
Samian ware, a name the propriety of which
has been disputed. The Roman writers speak

of an earthenware much used at table, and
said to have received its name from having
been originally made at Samos. It is described
as being of a red colour, as being of more
value than the common pottery, and as being
proverbial for its brittleness, all which characteristics belonged to the red ware found in
this country, which was covered with tasteful
subjects of all kinds in relief, and was evidently
much valued, as we often find vessels in this
ware which had been carefully mended, and
the brittleness of which was such that we
seldom find a specimen unbroken. Such
mendings, chiefly by means of metal rivets, are
exhibited in specimens of Samian ware found
in the excavations at Wroxeter, and deposited
in the Museum at Shrewsbury, where there
are also several pieces of this pottery, pre-
senting subjects which are interesting, and by
no means of common occurrence. It may be
further observed that the Samian ware in this
country resembles a Roman ware of which
the potteries have been found at Aretium, the
modern Arezzo, in Tuscany, but this ware was
much superior, especially in the degree of
artistic talent displayed in its ornamentation,
to that which was in use in this island, and
which no doubt was imported from Gaul,
where, especially on the banks of the Rhine,
the potteries in which it was made have been
found.

Extensive potteries have also been found
in this island, especially at Castor, in North-

amptonshire, where there was a Roman town
named Durobrivæ, and on the banks of the
Medway, at Upchurch, in Kent. The ware
from both these potteries is of a blue or slate
colour, produced by imperfect firing in what
is called the *smother-kiln;* that is, the air
being excluded and the heat being insufficient
thoroughly to bake the pottery, it retains so
much carbonaceous matter as to give it a black
colour. The pottery of these two establish-
ments is distinguished by the difference of
shapes. The ornamentation of the Upchurch
ware is in general of a very simple character;
that of the pottery from Castor is much more
elaborate, and often consists of hunting scenes
and other subjects, laid on in a white substance
after the pottery had been baked. Specimens
of both these wares are found at Wroxeter.

The excavations at Wroxeter have brought
to light at least two new classes of Roman
pottery, both evidently made in Shropshire.
The first is a white ware, made of what is
known as the Broseley clay, and consisting
chiefly of very elegantly formed jugs, with
narrow necks; mortaria, or vessels for rubbing
or pounding objects in cookery, the interior
surface of which is covered with grains of hard
stone; and bowls, which are often painted with
stripes of red and yellow. The other Romano-
Salopian pottery is a red ware, differing in
shade from the red Roman wares usually found,
and also made from one of the clays of the
Severn valley. Among the vessels in this

ware are bowls pierced all over with small
holes, so as to have served the purpose of
colanders. We find also some very curious
specimens of an imitation of the Samian ware;
but we have as yet no means of ascertaining
where it was made.

Many very interesting fragments of glass
vessels have also been found in the excavations
at Wroxeter. Two or three other objects
intended for domestic purposes have been met
with, such as a small bowl or cup made of
lead, and what appears to have been the handle
of some large vessel, made of block tin, neither
of which metals, used for such purpose, are
of common occurrence among Roman remains
in this country. A ladle and several knives
have also been found, and a handle of a knife
made of stone as well as several whet-stones.

Of personal ornaments the most numerous
are the hair-pins, most of which are made of
bone, though there are a few of bronze, and
one of wood. Their use was to hold together
the knot into which the Roman women rolled
up their hair behind the head, and through
which the pin was thrust. They are, on an
average, about three inches long, with a large
head, rudely ornamented ; and it will be
remarked that the shank is thicker in the
middle, and that it becomes generally thinner
near the head, no doubt to prevent the pin
from slipping out of the hair. Some of these
pins had evidently been saturated with an
oily substance, which shows that the ladies

F

in Roman Britain applied oil to their hair. Several fibulæ of the common Roman forms have been met with; they are all of bronze, of superior workmanship to the hair-pins, and most of those hitherto found at Wroxeter are, or have been, enamelled. Their use was to fasten the mantle and other parts of the clothing. Among the personal ornaments found already in the excavations are a number of buttons, finger-rings, bracelets, glass beads, and other objects, of which it is not necessary here to give a particular description. Of two combs, both of bone, one is remarkably neat in its form and make. Several bone needles may also be mentioned, and a pair of bronze tweezers for eradicating superfluous hairs.

Roman coins are found in considerable numbers, but many of them are so worn and defaced that it is no longer possible to decide to what emperor they belonged. The earliest met with during the present excavations is of the emperor Domitian. A great number are small coins of the Constantine family of emperors. Only two silver coins have yet been found, the others are of bronze or brass. The peasantry call them *dinders*, a name which, though it represents the Latin *denarius*, was no doubt derived from the Anglo-Norman *denier*.

Many objects of a more miscellaneous character have also been found during the present excavations, or have found their way into the Museum from former discoveries.

Among these are three artistes' pallettes, for
using colour; several weights, some marked
with Roman numerals; a steelyard; several
keys; portions of iron chains; styli, for writing
on wax tablets; an iron trident, which may
perhaps have been the head of a staff of office
or authority; one or two spear heads; a strigil
for scraping the skin in the sweating baths;
a portion of an iron horse-shoe; and two or
three very nice statuettes in bronze. The
most curious, however, of these miscellaneous
objects is a medicine stamp, intended to mark
packets or bottles of what, in modern times,
would be called patent medicines. A certain
number of these Roman medicine stamps have
been found in Britain and on the Continent,
and they are all, like this stamp, found at
Wroxeter, for salves or washes for the eyes,
diseases of the eyes having been apparently
very common among the inhabitants of the
western provinces of the Roman empire. The
Wroxeter stamp intended for a collyrium or salve
for the eyes, called *dialebanum* or *dialibanum*,
gives us in all probability the name of a
physician resident in Uriconium. The inscription may be read as follows, filling up the
abbreviations :—TIB*crii* CL*audii* M*edici* DIALIBA-
num AD OMNE VIT*ium* O*culorum* EX O*vo*, *i.e.*, the
dialebanum of Tiberius Claudius the physician,
for all complaints of the eyes to be used with
egg.

A few stones, with Roman inscriptions,
chiefly of a sepulchral character, have been

dug up at Wroxeter in the course of accidental excavations. Three of these were found in 1752, and are preserved in the library of Shrewsbury School. The first inscription may be read thus :—

C. MANNIVS
C . F . POL . SECV
NDVS . POLLEN
MIL . LEG . XX
ANORV . LII.
STIP . XXXI
BEN . LEG . PR
H . S . E.

intimating that it marked the grave of a soldier, of the twentieth legion, (which was stationed at Chester, the Roman Deva) named Caius Mannius, of the Pollian tribe. Another commemorated a soldier of the fourteenth legion, and has been supposed to belong to a very early period, as that legion was withdrawn from Britain before A.D. 68. It was the legion which suffered so much in the war against Boadicea, and this soldier may perhaps have been engaged in that war, although his having died in Britain does not necessarily imply that the legion to which he had belonged was there at the time, or indeed that it had ever been there, unless we had some other reasons for supposing that it had been there. His name was Marcus Petronius, the son of Lucius, of the Menenian

tribe, and the inscription may be read as
follows :—

```
        M . PETRONIVS.
        L . F . MEN
        VIC . ANN
        XXXVIII
        MIL . LEG
        XIIII . GEM
        MILITAVIT
        ANN . XVIII.
        SIGN . FVIT.         .
          H . S . E.
```

The third of these inscribed monuments
was divided into three columns or tables,
commemorating three members of the family
of a citizen of Uriconium, named Deuccus.
The inscription on the third column is entirely
erased, but the two others may be read as
follows :—

D . M	D . M
PLACIDA	DEVCCV
AN . LV	S . AN . XV
CVR . AG	CVR . AG
CONI . A	RATRE
XXX	

Another sepulchral stone, also preserved
in the library of Shrewsbury School was
found in 1810, and bore an inscription commemorative of Tiberius Claudius Terentius, a
soldier of the cohort of Thracian cavalry,
which may be read as follows :—

```
        TIB . CLAVD . TRE
        NITIVS . EQ . COH
        THRACVM . AN
        ORVM . VII . STIP
        ENDIORVM
          H . S.
```

In the excavation on the site of the
cemetery, in the autumn of 1862, a sepulchral
stone was found, which had not improbably
been placed over the door of a sepulchral
chamber of masonry. There has been a figure
above, the lower part of the legs and feet of
which alone remain. The slab bears the
following inscription, which from the damage
the stone has sustained is very difficult to
decipher, but I owe this reading to the knowledge and acuteness of my friend Mr. Roach
Smith. I may add that some of the letters
are extremely doubtful.

AMINIVS . T . POL . F . A
NORVMXXXXVSTIPXXII . MIL . LEG.
IIGEM . MILITAVITAQNVNC HIC SII
LEGITE . ET . FELICES . VITA . FLVS . MINV
IVSTAVINIERAQVATIEGIIIE . INTV
TANARA . DITIS . VIVITE . DVMSPI . . .
    VITAE . DAT . TEMPVS . HONESTE.

It is clear, at a glance, that the latter part of
this inscription contains three lines in hexameter
verse; unfortunately they are lines most rubbed
and most difficult to make out. Dr. McCaul,
president of the University of Toronto, in
Canada, in his recent work on " Britannio-
Romano Inscriptions," suggests that they may
be—

Perlegite et felices vitâ plus minus jutâ ;
Omnibus æqua lege iter est ad Tænara Ditis.
Vivite, dum Stygius vitæ dat tempus, honeste.

The two last words of the first lines are
extremely doubtful, and I confess that I do

not believe in Dr. McCaul's reading, which, of course, is but conjectural. The second does not appear at all to answer to what remains of the original, with the exception of the last words, Tænara Ditis. But of the last line, Mr. Smith's reading is much the best, and indeed appears to me to be the correct one,—

Vivite, dum spatium vitæ dat tempus, honeste.

The part preceding the verses may be read—

Aminius (perhaps Flaminius,) T*iti* P*ollione* F*ilius*, annorum XXXXV., stip*endiorum* XXII, mil*is* le*gionis* *v*ii gem*inæ*. Militavit aq*uilifer*. Nunc hic s*itus* *est*.

It may be remarked, that in many respects this is one of the most curious Roman inscriptions found in this island, and that it appears to be of rather an early date.

Another mere fragment of a stone, of the present existence of which I can learn nothing, is said to have contained the letters :—

LERT
FGAI
...TILES

Lastly, a monument of stone, which during the middle ages had been formed into a holy water stoop, and which is now in the vicarage garden, presents what has formed part of a Roman inscription—

BONA...REI
PVBLICÆ
NATVS.

It has probably been a dedication to one of the emperors, or an inscription commemorative of him.

IT has been stated before, that the site of
Uriconium is of very great extent. If the
visitor, after having examined the excavations,
would seek an agreeable walk, he may turn off
by the smith's shop already mentioned, along
the northernly continuation of the Watling
Street Road, which soon becomes a deep and
pretty country lane, and crosses the Bell Brook.
Soon afterwards, on the rise of a bank, we
come to a spot where the ancient town wall
crossed this road, and where there are said to
be traces of one of the gateway entrances to
Uriconium. At the latter part of the year
1862, excavations were made in an adjoining
field to trace the line of the town wall, which
was found remaining to a height of three or
four feet; but it was of very rough construction, built merely of small stone boulders
mixed with clay, and had evidently been
raised hurriedly, at a late period of the history
of Uriconium, to meet some sudden emergency.
There had evidently been an entrance opening
here, but there were no traces of gateway
buildings, which, perhaps, were only of timber.
Outside the walls, on the bank of the right,
was one of the principal cemeteries, and here
the sepulchral inscriptions mentioned above
were found. Successful excavations were made
in 1862 on the site of this cemetery, and
many Roman graves were opened, which fur-

nished the Museum in Shrewsbury with another inscribed monument of great interest, a number of sepulchral urns, and vessels of glass, and various other objects.

If, instead of going northward, the visitor follows the Watling Street Road towards the south, he will soon reach the village of Wroxeter, and may examine its church. A new gate to the churchyard has recently been erected, and Mr. W. H. Oateley, of Wroxeter, who holds the office of Churchwarden, has contributed a shaft of a Roman column, and two Roman capitals, which, together with another shaft given by the Rev. E. Egremont, are now placed on each side of this gateway. The two capitals which were dragged out of the river Severn are worthy of particular attention. They are singularly rich in ornament, and mark that late period of Roman architecture which became the model of the mediæval Byzantine and Romanesque. I cannot help wishing that they were safely deposited in the Museum at Shrewsbury, and I think that the Roman columns would serve as well for gateway supports without the capitals, which probably did not belong to them. The church of Wroxeter is a substantial Norman building, with later alterations, and on the outside of the southern wall of the chancel are the remains of a very interesting Norman doorway, which has been built up. The chancel internally is chiefly remarkable for some fine monuments with effigies of the sixteenth and beginning of the

seventeenth centuries, interesting especially
for their costume. It has at present a low
flat whitewashed ceiling, but there is a fine
old timber roof above, and it is greatly to be
regretted that the unsightly ceiling has not
been removed, so that the chancel might
again be open to its lofty roof of timber.
At the western end of the church is an early
font, *pl.* 7, which has been formed of a very
large Roman capital, taken from some important building in the city of Uriconium. Such
applications of Roman monuments to later
ecclesiastical purposes are by no means
uncommon. In the garden of the vicarage,
which adjoins the churchyard, are a few
fragments of Roman architecture and sculpture,
which have been carefully preserved by the
present vicar, the Rev. E. Egremont.

Near the churchyard stands the residence
of Mr. Oateley, who has also collected in his
garden a few fragments from the ancient city
and its neighbourhood. Among these is a
cylindrical stone, which at first sight might
be taken for a part of a column, but which
appears, from a few remaining letters of an
inscription, to have been more probably a
Roman milliarium or mile-stone. Mr. Oateley
has placed a Roman capital on the top of it,
and both are represented in *pl.* 6. Several
architectual fragments are also preserved in
the garden of the late Mr. Stanier. Two
of the most interesting of these, belonging
to the shaft of the same column, or to those

of two similar columns are represented in
pl. 15.

The Watling Street Road leads us direct
from the gateway of the churchyard to the
river Severn, which is here crossed at present
by a ford. On the right is a large rugged field
overlooking the river, and occupied by Mr.
Oateley, which has been trenched in several
directions, but nothing was discovered except
a Roman well, ten feet deep, which is kept
open, and is now partly filled with clear spring
water. In an orchard at the corner of this
field, near the road, were found a number
of human skeletons, attended with some
remarkable circumstances, for an account of
which I refer the reader to Dr. Johnson's
remarks at the end of this little volume. On
the other side of the Watling Street Road,
the ground rises to a little knoll, which looks
down upon the river, and seems to have
formed the southern corner of the inclosure
of the city of Uriconium. The top of this
knoll has been carefully explored, and the
walls of a square building, perhaps of a tower,
were uncovered. Among the objects found
on this spot were a head sculptured in stone,
and a mould for casting Roman coins, both
of which are deposited in the Museum at
Shrewsbury. The impress on the coin-mould
is that of a coin of Julia Domna, the wife of
the emperor Severus, (the commencement of
the third century); and it is rather a curious
circumstance that a silver coin of this empress,
which fits the impress exactly, has been found

in the excavations near the Old Wall. This method of multiplying the imperial coinage by casts seems to have been very common in these distant provinces, and was perhaps exercised by the imperial or municipal officers. Another coin-mould, also with the impress of Julia Domna, was found at Wroxeter in 1747, and two, one of Severus himself, and the other of Plautilla, in 1722.

———

In conclusion, I may perhaps be allowed to make a remark on some of the various points on which the excavations on the site of Uriconium have already thrown more or less illustrative light during the short period in which they have as yet been carried on. We see how, by examining their buildings and comparing the objects that are turned up by the pick and the spade, we get an insight into the condition of the inhabitants of Roman Britain, and learn to what degree they enjoyed the luxuries and comforts of life. We see that they possessed a great majority of the refinements of modern society—far more than can be traced among the population of the middle ages. We are taught even the character of their food by remains of edible animals. The comparison of other objects enables us to judge to a great degree of the state and extent of manufactures and commerce. We learn from inscriptions on their sepulchral monuments and altars the names and occupations of some of the inhabitants of the ancient town, and the races to which they belonged; and from this partial information we are enabled by induction to obtain a general view of the whole. We are thus enabled to form a truer notion of the manner in which this country had been inhabited and governed during nearly four centuries; and we have the further hope of eventually discovering monuments which will throw some light on the more particular history of this neighbourhood in these remote ages. We learn, finally, from the condition in which the ruins of Uriconium are now seen, and especially from the numerous remains of human beings which are found scattered over its long-deserted floors, the sad fate under which it finally sank into ruin, and thus we are made vividly acquainted with the character and events of a period of history which has hitherto been but dimly seen through the vague traditions of writers who at best knew them only by hearsay.

Catalogue of Wroxeter Antiquities in the Museum at Shrewsbury.

—

I.

OBJECTS CONNECTED WITH THE ARRANGEMENT AND CONSTRUCTION OF THE HOUSES AND OTHER BUILDINGS.

1. Roofing flags, of micaceous sandstone, form generally hexagonal, with a hole for the nail.—*See pl.* iv., *fig.* 1.

2. Tiles of various kinds :—small square tiles, flue-tiles, roof-tiles, &c. Large oblong square tiles for bonding-courses in the wall, &c. Square tiles for making the pillars in the hypocausts.

3. Specimens of the concrete which covered the hypocausts, to the depth of eight inches or more, forming the floor of the apartment.

4. Eight different specimens of the tessellated, or mosaic pavement, taken up as it was found, and framed.

5. Drawings of the same, made by Mr. George Maw, of Broseley, and presented by that gentleman to the Museum.

6. Sculpture in sandstone; a head, of late Roman art, which appears to have formed part of the architectural ornamentation of a building.

7. Bases, capitals, and shafts of columns.

8. Stucco, covering the walls, coloured, plain, and with some formed patterns. One specimen, bearing the letters A.R.C.A., having formed part of an inscription on the wall. Tessellated ornamentation of the surface of a wall, dark and light tesserae, so as to form an irregular pattern.

9. *Umbilicus*, or hinge for a door.

10. Iron bolts, T shaped iron stancheons, and nails, for fixing roof and flue-tiles upon the walls.

11. Many tiles bearing the impression of the foot of domestic or wild animals,—some of the dog; others of sheep, pig, horse, and ox.

II.

OBJECTS FOR DOMESTIC PURPOSES.

1. POTTERY :—

a. Samian ware.

b. Upchurch pottery.

c. Durobrivian pottery.

d. Romano-Salopian ware, made of clay obtained from Broseley.

e. Pieces of red earthenware, probably made in Shropshire.

2. GLASS :—

a. Fragments of flat or window glass.

b. Portions of bottles, &c., generally coloured, some opalescent.

c. Fragments of a cup, ornamented with spots of deep purple glass.

d. One green glass bottle, 6½ inches high, with narrow neck, found in the cemetery, quite entire.

e. A green glass jar, with wide mouth, about 5 inches high and 6 inches wide, also found in the cemetery. It was full of soil, everywhere penetrated by roots of plants.

f. Two metallic mirrors or *specula*, one in fragments, the other entire. They are of white metal, a compound of tin and copper, with a large proportion of the former.— (Cemetery.)

g. Three very pretty lamps. One bears the figure of Hercules, another that of a dolphin, a third that of a boy kneeling.— (Cemetery.)

h. A silver fibula.

3. Bronze statuette of Venus and Mercury.

4. A *strigil*, (fragment.)

5. Part of an iron horse-shoe. Iron bit of a bridle. Iron spur.

6. Two masks, one made of terra cotta, the other of pottery.

7. Anomalous earthenware vessel.

III.

IMPLEMENTS AND UTENSILS.

1. Weights: one in lead, 20¼oz.; another in stone, 11½oz.

THE RUINS OF URICONIUM.   89

Weight in lead, marked ii., weighs 2¼oz.;
another also in lead, weighs 2½oz.
    2. Ladle; and neck of some vessel made
of block tin.
    3. Several keys, of different forms. Iron
padlock.
    4. Large shakels, chains, &c., of iron.
    5. Knives, spear-heads, and portions of
other weapons. Two axe-heads. Bone handle
of a sword, *very curious.*
    6. Several whetstones. Stone handle to
a knife. Touchstone.
    7. Iron trident, supposed to be a candle-
stick.
    8. Rings of iron, bronze, and lead.
    9. *Styli* of bronze and iron; bronze
tweezers; bronze and iron spoons; steelyard.
    10. Small cup of lead; ditto of thin copper.
    11. Large plates of lead, purpose unknown.
    12. Cock made of lead, a child's toy.
    13. Fragment of a lamp in red pottery.
    14. Three painters' palettes.
    15. A curious iron box—(ointment box.)
    16. Iron trowel.
    17. Bronze lancet (?)
    18. Bronze pan of a pair of scales.

IV.

PERSONAL ORNAMENTS.

    1. Hair pins, in great variety; more than
30 specimens have been found made usually of
  G

bone; with some of bronze, but those are much
more slender. *Pl.* 11.

2. Bodkins or needles made of bone.
3. Fibulæ, and buckles in great variety.
4. Bracelets or armlets, and brooch.
5. Bronze studs or buttons, some flat and
others very convex.
6. Finger rings:—*a.* silver; *b.* yellow
bronze; *c.* bronze, with iron wire; *d.* bronze,
with open work on one side; *e.* fragment of one
of wood; *f.* iron signet ring—device engraved
upon a blue stone, a fawn coming out of a
nautilus shell.
7. Combs made of bone, one much ornamented. *Pl.* 10, *figs.* 5, 6.
8. Beads of glass of various sizes, some
large to suspend round the neck, others to
string together upon a thread.
9. Bronze bracelet of twisted work.
10. Amber finger ring, (found 1867.)

COINS.

1. Coins found in the present excavations
at Wroxeter.
2. Coins found at Wroxeter, at different
times, and given to the Museum.
3. The coins found with a skeleton in
the hypocaust.
4. Coining-mould of baked clay. Julia
Domna.

CINERARY URNS.

1. Large red earthenware urn, containing

human bones (burnt), inclosed in an outer urn of lead, which was brought from Wroxeter many years ago.

2. Another Cinerary urn of black pottery, cotaining burnt human bones, found in a field adjoining the .Cemetery, and outside the town walls. Purchased by the secretary.

3. A large Cinerary urn, found in the recent excavations, ten inches high, and thirty in circumference, almost entire, containing bones, but not human.—*See pl.* 13, *fig.* 2.

4. Cinerary urns in red and black pottery of various sizes, from 4 to 12 inches high. Some containing burnt human bones and unguent bottles.—(Cemetery.) Many small flask-shaped bottles were found, some broken, some entire, some which had evidently been exposed to heat. Oily matter was detected in one ; hence they have been termed unguent bottles.— (Cemetery.)

V.

MISCELLANEOUS OBJECTS.

Medicine Stamp, found at Wroxeter, in 1808, by Mr. Upton ; purchased from his family in 1859, by the late Beriah Botfield, Esq., M.P., who presented it to the Museum.

1. Oyster shells in great number; shells of some nut found in an oyster shell.

2. Remains of small animals and birds.

3. Nondescript articles in iron, shapeless masses of lead, innumerable fragments of pottery, bone, &c.

4. Fragments of horn and bone, which have been cut with a saw or other tool.

5. Fragments of bone, which have been turned in a lathe.

6. Inscribed sepulchral stone with Latin inscription, partly legible. There has been a statue on the top.—(Cemetery.)˙

7. A skiff-shaped vessel in bronze, with round handle, and a lid which closed with a catch.

8. Several legs of the fighting cock, with very large natural spurs.

9. Roundels, formed chiefly from the bottoms of earthenware vessels, perhaps used in some game; others made with a hole in the centre.

10. Skulls of the dog: one, that of a dog of the mastiff kind, of an unknown species. Bones of horse, ox, roe, and red deer, (*Cervus elaphus*); also fragments of the horn of a species allied to the elk of Ireland. (*Strongylocerus spelæus.*) Very numerous remains of the wild boar, including bones of the hoof, jaw, and tusks.

Among other bones of the ox are some of a very large kind, now unknown in this country.

Also, the crania of the *Bos longifrons*, more than one bearing evident marks of the fatal blow of the axe on the forehead.

11. Specimen of *mended* pottery:—1. Samian ware. 2. Upchurch. 3. Romano-British pottery.

12. Slabs of stone for grinding or mixing colours, painters' palettes.

13. Specimens of hepatic iron ore. Ditto of barytes, or heavy spar.

14. Iron tire of a wheel, 3ft. 3in. in diameter, 1½ inch in breadth. Two iron hoops, supposed to have belonged to the nave of the same wheel.

15. Two hoops of another nave, with the wood remaining between them.

Human Remains.

1. PARTS of three human skeletons found in the hypocaust B. Two of the skulls are almost entire, and one is broken into fragments. The latter is remarkable for its great thickness. One of the two former, from its form, is most probably the head of a female, and the bones of the pelvis of one skeleton are also characteristic of the female sex. One jaw-bone must have belonged to a very old person, as not only the teeth but even the sockets are gone. One hundred and thirty-two coins were found in the hypocaust with these skeletons. See page 39.

2. Five human heads, and other parts of human skeletons, were first dug up in the orchard, near the river. Of these, *four* were singularly deformed,—the one eye being in advance of the other, and the face oblique.

Ten other skulls have since been found in the
same place, and have been arranged in the
Museum. Of the ten above-mentioned *three*
are deformed like the others, four are so broken
and defective that their form cannot be ascertained; three are not deformed. One of the
latter is a very large skull, well formed, but
with very strong projecting cheek (malar) bones,
and a projecting occiput.

3. The principal bones of a skeleton
(female?) belonging to one of the skulls,
stretched on a board (as well as could be done
on the spot) just as it lay in the ground.

The circumstances under which these
skeletons were found are full of interest. The
greater part of them (at least twenty have been
found, but not all in a state to be taken up)
were evidently put into the ground with a
certain degree of attention, that is, *buried*. They
were not thrown heedlessly into a pit, but carefully deposited at full length, and generally
near together, the legs and arms for the most
part extended, or, as in the case above described,
one arm lying across the body. In general,
nothing particular has been found near them,
but only the usual contents of the soil, such as
stones, roots, and fragments of pottery. In one
instance an iron ring, in another some nails
were met with, and in a third a single coin of
Claudius Gothicus. But all these might have
occurred accidentally in the neighbourhood of
the bodies, in an old Roman site, and not have
been buried with them. No vestiges of wood

derived from coffins, or of apparel, were discovered. There were no traces of weapons or articles of domestic use, which were generally buried by the Romans with their friends, and the place where these remains were found is within the walls, and could not, therefore, be a Roman cemetery.

4. In more than one instance, bones of very young children have been found; but in one instance, alluded to at page 65, almost an entire skeleton of a child was found, which has been preserved, and is in the Museum. This was found outside the semi-circular end of the great hypocaust, where there must have been a small court. From the smallness of the bones of the skeleton, and from the circumstance of the teeth being still contained within the jaw-bone, it may be inferred that this was a very young infant—perhaps still in arms.

5. A thigh bone has been found, which, having been fractured, has become united during life.

The most interesting circumstance connected with the human remains found at Wroxeter is the large relative proportion of deformed skulls. Of the nineteen crania found in the orchard and since deposited in the Museum, eleven are more or less crooked. It has been supposed, and indeed the opinion is still entertained by some antiquarians, that this deformity was *congenital* and not *posthumous*, that is to say, that the persons to whom these skulls belonged lived and died with deformed heads. And this was my own view before I had learned that bones are capable of being bent by pressure in the ground. There can be little doubt that the deformity has been produced by posthumous pressure, aided by moisture and the solvent action of certain acids that always exist in vegetable mould.* Other instances of a like effect have been described by Dr. Sherman,† and, in America, by the Rev. D. Wilson.‡

H. J.

---

* See Abstract of Proceedings of Royal Society, June, 1862.
† Crania Britannia.
‡ Pre-Historic Man, vol. ii., p. 306, &c. &

APPENDIX.

———

On November the 11th, 1858, at the General Meeting of the Shropshire and North Wales Natural History and Antiquarian Society, held at the Museum, Shrewsbury, the President, Beriah Botfield, Esq., M.P., proposed that excavations should be commenced at Wroxeter. He had written to the Duke of Cleveland, and obtained his Grace's consent to do so. He also made the very liberal offer to give fifty guineas towards the expenses, provided that fifty other gentlemen could be found willing to subscribe one guinea each. A Committee was formed, consisting of the following noblemen and gentlemen to carry on the work:—

The Right Hon. the Earl of Powis, Powis Castle
Beriah Botfield, Esq., M.P., F.R.S., Decker Hill
R. A. Slaney, Esq., M.P., Walford Manor
Rev. B. H. Kennedy, D.D., Shrewsbury
Rev. E. Egremont, Wroxeter
Rev. R. W. Eyton, Ruyton, Shiffnal
Rev. H. M. Scarth, Bathwick
Samuel Ashdown, Esq., Uppington
W. H. Bayley, Esq., Shrewsbury
William F. F. Ffoulkes, Esq., Stanley Place, Chester
Henry Johnson, Esq., M.D., Hon. Sec., Shrewsbury
George Stanton, Esq., Shrewsbury
Albert Way, Esq., Worham Manor
Samuel Wood, Esq., Shrewsbury
Thomas Wright, Esq., F.S.A., Brompton

A Metropolitan Committee has since been thought desirable, and held its first meeting August 3rd. It consists of the following distinguished noblemen and gentlemen:—

The Right Hon. Earl Stanhope, President of the Royal Society
of Antiquaries
The Right Hon. Viscount Hill, Lord Lieutenant of Shropshire
The Right Hon. Lord Braybrooke
The Right Hon. Lord Talbot de Malahide
The Right Hon. Lord Lindsay
The Right Hon. Lord Newport. M.P.
The Right Hon. the Lord Chief Baron
Beriah Botfield. Esq., M.P.
The Hon. Rowland C. Hill. M.P.
R. Moncton Milnes, Esq., M.P.
C. Octavius S. Morgan. Esq., M.P.
H. Danby Seymour, Esq., M.P.
W. Tite. Esq., M.P.
C. C. Babington, Esq., F.R.S., St. John's Coll., Cambridge
The Rev. E. L. Barnwell, General Secretary of the Cambrian
Archæological Association.
Sir John P. Boileau, Bart., F.R.S., V.P.S.A.
The Rev. Dr. Bosworth, F.R.S., F.S.A., Professor of Anglo-
Saxon, Oxford.
The Rev. Dr. J. Collingwood Bruce, F.S.A., Hon. Sec. of the
Society of Antiquaries of Newcastle upon-Tyne
Talbot Bury, Esq., F.R.I.B.A., A.I C.E.
Benjamin Bond Cabbell, Esq., F.R.S., F.S.A.
Robert Chambers, Esq., Edinburgh
Sir James Clarke. Bart., F.R.S.
James Dearden. Esq., F.S.A.
C. Wentworth Dilke, Esq.
J. Hepworth Dixon, Esq., F.S.A.
Joseph Durham. Esq., F.S.A.
The Rev. E. Egremont, Vicar of Wroxeter
F. W. Fairholt. Esq., F.S.A.
Augustus Guest, Esq., LL.D., F.S.A.
S. Carter Hall, Esq., F.S.A.
J. O. Halliwell, Esq., F S.A., F.RS.
The Rev. C. H. Hartshorne.
Fredk. Hindmarsh, Esq., F.R.G.S., F.G.S., Hon. Sec.
The Rev. T. Hugo, F.S.A.
Dr. Henry Johnson, Hon. Sec. of the Excavation Committee.
Shrewsbury
Joseph Mayer. Esq., F.S.A., Liverpool
Sir Roderick I. Murchison. F.R.S., &c.
Frederick Ouvry. Esq., F.S.A.
The Rev. H. M. Scarth
Charles Roach Smith, Esq., F.S.A.
Vice-Admiral W. H. Smyth, F.R.S., F.S.A.
W. S. W. Vaux, Esq., F.S.A., President of the Numismatic Society.
Albert Way, Esq., F.S A.
Thomas Wright, Esq., F.S.A., Treasurer.

In a handsome volume, in royal 8vo., with numerous illustrations, price 25s.,

URICONIUM;

THE HISTORY AND ANTIQUITIES OF THIS ANCIENT

BURIED CITY,

AND OF THE

Romans in Shropshire.

BY THOMAS WRIGHT, ESQ.,

M.A , F.S., HON. F.R.S L., &c.,

Corresponding Member of the Imperial Institute of France, (Académie des Inscriptions et Belles Lettres.)

THE above work contains a complete account of the researches which have been hitherto made on this interesting site, illustrated and explained by a comparison with the similar antiquities found on different sites in Britain, as well as in other parts of the Roman Empire. It contains a history, as far as it can be traced from existing materials, of the Roman occupation of this part of the island; a complete account of the discoveries which have been made on the site of the city of Uriconium; and an attempt to display, by means of these, the condition, life, and manners, of the Roman inhabitants of this island.

BUNNY & EVANS, Printers, High Street, Shrewsbury.

The Shrewsbury Chronicle,

— AND

Shropshire and Montgomeryshire Times,

(Established 1772.)

THE COUNTY NEWSPAPER, AND LEADING JOURNAL FOR SHROPSHIRE AND NORTH WALES,

has by far the LARGEST CIRCULATION of any Newspaper published in the district, its Guaranteed Sale being FIVE TIMES more than any other Shrewsbury Paper, spreading over upwards of 100 towns and villages, in most of which it has established Agents and Correspondents. It is

THE GREAT ADVERTISING MEDIUM THROUGHOUT SHROPSHIRE AND THE PRINCIPALITY.

PUBLISHED EVERY FRIDAY MORNING.

Price 2d. By Post 2½d.

JOHN WATTON, Proprietor.

SHREWSBURY.

THE
RAVEN FAMILY HOTEL,
NEAREST TO THE RAILWAY STATION.

The number of Bedrooms having been largely increased, Visitors who have been disappointed hitherto may now rely upon obtaining comfortable accommodation.

BILLIARDS, POSTING, &c.

Special arrangements made with Families and Gentlemen to board by the week or month.

MISS COLLETT, Manager.

T. L. D. JONES'S
Shrewsbury Horse Repository,

And the Shropshire & West-Midland Counties Horse Mart, Established 1867, for the Sale and Purchase of Horses of every class.

AUCTION SALES, EVERY ALTERNATE TUESDAY, throughout the year. and always a lot of Gentlemen's Hacks and Harness Horses on Private Sale.

FIRST-CLASS POSTING ESTABLISHMENT.

ELEGANT WEDDING EQUIPAGES, Broughams, & Flys, Summer Carriages, Dog Carts, &c., Job Horses, Ladies' Pads, and Gentlemen's Saddle Horses.

LIVERY & BAIT STABLES.—Scales of Charges and Terms, which are most moderate, are posted about the premises, and free by Post.

THE AUCTION ENTRY FEE ON HORSES IS REDUCED TO 5s. EACH.—EARLY ENTRIES for each Sale are respectfully solicit d to be made in time for publication in the previous week's newspapers, and in the Catalogues issued four days previously.

T. L. D. JONES,
Auctioneer and Proprietor, and
the Shropshire Horse Master's Agent,
The Repository, Shrewsbury.

Eddowes's

Shrewsbury Journal,

AND SALOPIAN JOURNAL,

(Established 1794,*)*

PUBLISHED EVERY WEDNESDAY MORNING,

PRICE TWOPENCE.

EDITOR, MANAGER, AND PUBLISHER,

JAMES SHARPE.

THE JOURNAL has a large and influential circulation throughout the County of Salop and the whole Principality of Wales, and also an advertising patronage amongst Capitalists, Solicitors, Auctioneers, Merchants, Land Agents, and Traders, SUPERIOR TO THAT OF ANY OTHER NEWSPAPER published in the District. It also circulates extensively in the neighbouring Counties, and will be found at the principal Hotels and Commercial Offices in London, Birmingham, Liverpool, Manchester, and other important towns. It is thus UNQUESTIONABLY THE BEST MEDIUM FOR ADVERTISING, and affords a safe and widely-spread means of publicity amongst all those classes most likely to be useful to Advertisers.

OFFICES :—7, THE SQUARE, SHREWSBURY.